I0044866

Reverse Engineering the Universe

Using One Particle and Three Forces

"I, at any rate, am convinced that He (God) does not throw dice."

—*Albert Einstein in a letter to Max Born*

Reverse Engineering the Universe

Using One Particle and Three Forces

This book clarifies and expands on ideas originally proposed in
I Killed Schrödinger's Cat: A Simple Theory of Energy and Matter in the Known Universe by Donald A. Bertke and Herbert L. Hirsch, 2014

Donald A. Bertke
Herbert L. Hirsch

Cover illustration by Susan A. Bertke

Bertke Publications
P.O. Box 291974, 1740 East Stroop Road
Dayton, OH 45429-9998
U.S.A.

2017

Reverse Engineering the Universe: Using One Particle and Three Forces

Copyright © 2017 by Donald A. Bertke and Herbert L. Hirsch

All rights reserved. This book or any portion thereof may not be reproduced or used in any manner whatsoever without the express written permission of the publisher except for the use of brief quotations in a book review or scholarly journal.

First Printing: 2017

ISBN 978-1-937470-19-7 Paperback
ISBN 978-1-937470-20-3 PDF

Library of Congress Cataloging-in-Publication Data

1. Physics. I. Title

Bertke Publications
P.O. Box 291974, 1740 East Stroop Road
Dayton, OH 45429-9998

bertkepubs@email.com

Table of Contents

Preface

TON (The Only Needed) Particle Theory evolved out of my frustration with quantum mechanics and string theory, which both create more mathematical complexity without adding to our understanding of the universe. As a result of this frustration, I went back to the origins of Einstein's equation, $E=mc^2$, which relates mass (m), energy (E), and the speed of light (c). After taking a clean look at the formula's implications, I discovered that the classic interpretation is WRONG! You cannot create energy by destroying or transmuting mass, nor can you convert energy into mass. Yet, I see intelligent scientists repeat this absurdity day after day.

As I approached the universe with an engineering perspective, I discovered a solution whose simplicity is worthy of consideration. TON Particle Theory brings the universe into clarity with just four dimensions and three basic and well documented forces that can explain everything from dark matter and energy to the Big Bang (the universe expansion), black holes (Large Dense Mass Objects (LDMO)), and the eventual end of the current universe (the universe collapse) and its rebirth (the next universe expansion).

So, open your mind and take a look at this theory. When you are done understanding its premise, I ask you to prove it wrong. That pursuit should prove to you that TON Particle Theory is correct and is the answer to many questions we have about the universe.

—Donald A Bertke, Beavercreek, OH

Having known Don and worked with him for several decades, I always expect two things from any of his ideas. The first is creative thinking, and the second is soundness. Another way of putting this is that, although whatever he comes up with may seem strange or unconventional, it usually turns out to be true.

As a practicing engineer and scientist over the past four decades, I have seen almost invariably that whenever someone concocts an elaborate solution to a problem or a complicated explanation of a phenomenon, they have overlooked the true and simple basis for it: they have not dug deep enough or measured precisely enough. There are myriad instances in the history of scientific progress where a new theory or discovery was only possible when its discoverers had the tools or understanding to look further beneath the surface.

Hence, when Don first started talking to me about "TONs," his sound and simple explanation of things was very scientifically appealing because it *did* look beneath the present "surface" and stuck to rational scientific bases from pioneers such as Einstein, Newton, Coulomb, Maxwell and others. It had always seemed to me that many aspects of the subatomic particle theory we have today were over-complicated attempts to mathematically rationalize one theory or another, because application of fundamentals did not seem to explain things. Folks were searching far and wide for explanations that were actually close but deep. TON Particle Theory *is* the simple explanation for all things, which has so long evaded scientific thought and practice.

A final point is that the genesis of what appears in this publication is all Don's original ideas. My part was working out some of the mathematics, debating ideas and concepts as they emerged, and perhaps moving some ideas along now and then. I was happy to be included as a co-author.

In our first edition, we included an Appendix containing three mathematical proofs to support the premises asserted in the book's main body. These proofs were designed to verify, to the extent possible, the plausibility of some of TON-to-TON and atomic level force relationships and dynamics. In this second edition, we added two more mathematical proofs: (1) to offer additional insight into plausible attractive and repulsive force balancing and (2) to describe a simpler atomic structure using the components of TON Particle Theory.

—Herbert L. Hirsch, Vandalia, OH

As a beta reader and editor for Don and Herb, I acted as their chief guinea pig. I've never taken a formal physics class, and my math education ended with trigonometry. Don's not-very-flattering idea was that, if *I* could understand what they were saying, anyone could.

I very quickly discovered that just about everything I had learned about atoms in school and from general reading and television programs was diametrically *opposed* to what Don was saying. As just one very quick example, in TON Particle Theory there are neither electrons nor neutrons in atoms. *Heresy!*

I therefore offer the following advice: If you are a student who wants to get a good grade in science, do *not* use anything you read this book, and certainly do *not* quote TON Particle Theory in your exams. In fact, it's probably best not to even *read* this book until you've graduated and may think for yourself.

Seriously.

—S. A. Bertke

Introduction

When engineers first encounter an unknown system, they begin a process called "reverse engineering," by which they take measurements and make observations, and then analyze that information to explain how that system might work. When viewed as a system, we can do the same with the universe.

Throughout the ages, many scientists have collected a wealth of information using various scientific instruments. One of the goals of these observations and measurements was to understand how the universe may have developed into its present form and how it functions. Scientists analyzed the collected data and tried to establish a logical process by which the universe could have emerged and evolved from the beginning to our present day. During the analysis process, scientists and engineers developed mathematical models that proposed how the intricate and natural processes of the universe could function. They then used these mathematical models to predict the next logical, evolutionary stages of the universe.

This book provides the reader with our proposed models and their derivation, along with valid mathematical bases for how and why the universe could work the way we describe. We are confident that TON Particle Theory provides valid answers based upon actual, observed, physical phenomena, instead of the purely mathematical conjectures proposed by other leading theories.

In addition, TON Particle Theory provides a simpler explanation for the scientifically "explained" phenomena ascribed to the current universe. Our approach, using simple constructs, redefines the basic components of the universe. When combined with our simple stellar fusion model, TON Particle Theory easily explains all currently observed and measured phenomena. If science has taught us anything, it is that simpler is better.

True enlightenment can only occur after you realize that every-thing you have been taught is mostly incorrect. At that point, true learning and understanding can begin.

Chapter 1. TON Particle Theory Overview

This book describes a series of thought experiments, analyses, and investigations into what we call TON (The Only Needed) particles or "TONs" (pronounced "tahns"): their underlying basis, what they are, and how they interact to form all the "things" in our universe—from subatomic particles to so-called black holes. During our research (reviewing experimental data and assessing scientific papers), we looked in many directions, had some enlightening moments or "epiphanies," and, at the end, found ourselves with a basis for a whole new approach to subatomic particle theory and its implications to our understanding of the universe.

In this overview, we introduce the basic concepts, and in the ensuing chapters we give greater detail about our investigations and their results.

Reinterpreting Einstein's Equation

In our research, we concluded there is a major problem with our current understanding of what Einstein's most famous equation really says. Let's start there:

$$^1 E = mc^2 \text{ or more generally } E = mv^2 \quad (\text{Eq.}1)$$

Where: E is Energy,
m is Mass,
c is the speed of light, a specific velocity[2], and
v is velocity of the moving mass, m.

---

[1] Sir Isaac Newton defined kinetic energy to be ½ mv², which brings up an issue of why Einstein failed to include the ½ factor in his equation. If viewed as the kinetic energy of a mass object at its maximum velocity, a case can be made for its omission.

[2] In later analysis, we concluded that the term "speed of light" is actually the fastest rate at which a captured photon can be either absorbed or emitted from a photon shell. You may consider it as the event rate limit of the universe.

Many scientists have interpreted this equation to say that mass and energy are transposable or interchangeable. The most repeated phrase we hear in some scientific circles is that we can create energy from mass and mass from energy. While interesting, that view is incorrect.

The equation clearly says that the maximum kinetic energy (E) is obtained when we have a mass object (m) traveling at the speed of light (c). It simply defines the energy of a moving mass or, as Sir Isaac Newton defined it, kinetic energy. The equation also implies that the maximum energy a given mass can carry is attained at the speed of light. That interpretation is very different from creating energy from mass, or vice versa.

All Energy Requires Mass.

If Einstein's equation is correct, simple algebra tells us it is impossible to have massless (quanta[3]) of energy. Assigning a value of zero to m, mass, you get:

$$E = 0c^2 = 0 \quad \text{(from Eq.1)}$$

So, if you have zero mass, you have zero energy. This simple conclusion affects our understanding of the universal phenomenon we call light.

The First Problem with Light

In his paper on the photo-electric effect, Einstein demonstrated that light has energy. Yet, because the mass of light could not be directly measured, physicists of the time assumed it was massless. This assumption about the nature of light created a logical problem, since scientists have confirmed that light (a photon) passing near a large, dense mass object (LDMO) such as a star alters its path in space. (This is called the gravitational lensing effect.) As Sir Isaac Newton's work on gravity tells us, this lensing effect should not be possible in a universe where photons have no mass.

To explain how a LDMO could alter the path of massless light,

---

[3] The term "quanta" was introduced by physicists to explain energy phenomena that appeared to have no mass, but appeared to occur at specific integral instances.

Einstein proposed that gravity creates a distortion in the fabric of space (a gravity well) that distorts the path of light. So, it is not the *light* that is bending, but space itself. He did not believe that gravity attracted light (photons), as Newton's gravitational formulae described.

Instead of concocting theories to explain why the universe is contradicting our data, let's change our basic assumption to:

Light Has Mass

TON Particle Theory explicitly states that light (photons) have mass. Having mass, no matter how slight, photons respond normally to Newton's gravitational formulae.[4] With the change in this single assumption, the conflict between theory (massless photons) and observation (photons apparently bending due to a strong gravitational field) disappears.

The Second Problem with Light

Scientific observations clearly document that photons behave like particles but maintain their frequency of vibration like a wave. This is true at the frequencies of "visible light" as well as throughout the full gamut of electromagnetic phenomena. Furthermore, when performing the "two-slit experiment," light appears to behave as a wave since it appears to recombine into a single wave front after passing through the slits.[5] This duality in behavior led to the ongoing debate about the true properties of light/photons. Are photons (supposedly massless) particles or waves of pure energy?

Since we have unambiguously stated our position that photons have mass, TON Particle Theory is firmly on the "photons are particles" side of the debate. How, then, do we explain their wave-like behavior?

First, all forms of photons vibrate at specific measured frequencies or wavelengths to encompass the full spectrum of electromagnetic

---

[4] Our estimates for the mass and size of the smallest possible (unit) photon are provided in Chapter 2.

[5] For more information on this experiment, search the Internet for double-slit or two-slit experiment.

phenomena.[6] To get such sustained vibration, physics requires a *pair* of forces to repel and attract *two* mass objects of similar or equal mass without energy loss. A single mass object cannot inherently vibrate, because it requires a second mass to work in conjunction with the two forces to satisfy Newton's Third Law: *For every action, there is an equal and opposite reaction.*

A Single Mass Particle Cannot Vibrate.

Forget what you've been taught for a moment and focus on the following logic chain:

- You must have two mass objects to have vibration.
- Photons vibrate.
- Therefore, photons must be composed of two mass objects.

> From now on...
> ## PHOTON =
> *Two TON particle objects of similar mass and charge that vibrate or rotate in a gravitational and electrostatic entanglement.*

In Chapter 2, we show that the smallest possible photon (unit photon) is indeed composed of two very small mass objects, which we call TON particles or TONs, vibrating against each other. The specific vibration frequency of any photon depends on the mass and charge of the two TON particle objects within it.[7]

So, photons are the source of all electromagnetic phenomena[8] and possess the kinetic energy that science currently knows and uses.

What about the wave effect?

The reason that photons produce apparent wave fronts comes from the gravitational lensing effect of the two-slit experiment material and the lack of a fully collimated illumination source. When a photon

---

[6] See Chapter 2, where we explain the stellar process by which photons are created and how the process creates photons of various masses that vibrate from incredibly high frequencies to extremely low frequencies.

[7] The math for the vibration/oscillation of the photon is given in Proof 1 in the Appendix.

[8] One caveat—At some point the electromagnetic energy begins to move mass objects as surface waves instead of relying upon an actual photon configuration. We are not yet sure of when this transition takes place. We suspect that it lies in the Terahertz range.

passes through the slit aperture, the mass of the slit material establishes a brief gravitational pull on each photon as it goes through the slit. Only a photon that passes perfectly through the exact center of the aperture can remain on its straight-line trajectory. All other photons will experience a slight tug towards the closest aperture edge. As a result of these forces, the various photons will fan out from the aperture in a spherical trajectory, which would resemble a wave front. Problem solved.

What about the fringe patterns?

Again, gravity causes the photons to deviate as they pass through the aperture. Photons have two mass objects inside. Since there are many possible orientations for these TON particle object pairs within the photon, the photon trajectory will be slightly different depending upon the internal TON particle objects' orientation as they go through the aperture. Therefore, those photons with similar TON particle object orientation will tend to be deviated by the same amount, while other orientations will be affected differently. In the far field, you will see them separate into bands based upon the predominant aperture orientation and the orientation of the individual photons as they entered into the gravity effect of the aperture.

What is mass made of?

In classic physics, all mass is composed of atoms, which are generally described as nuclei surrounded by orbiting electrons. The nucleus, in turn, is composed of two basic subatomic particles: the proton and the neutron. There is a growing list of various sub-subatomic components which scientists theorize exist inside the proton, e.g., Fermions, quarks, leptons, bosons, etc., but nobody can definitively claim to have seen, measured, or otherwise established what goes on inside atoms and their basic building blocks.

TON Particle Theory proposes that, instead of protons, neutrons, and electrons, the universe has *one basic particle*. As mentioned earlier, we call this particle a **TON** (**T**he **O**nly **N**eeded) **Particle**. From readily available scientific data, we can deduce a great deal about the properties this particle must have:

- **Spherical Form.** A sphere represents the most efficient geometry for an object. A spherical particle best supports the measured and observed characteristics of TON particles for mass and charge, both key components of TON Particle Theory.

- **Incompressibility.** The basic particle cannot be made smaller or distorted into a non-spherical shape. There is a physical limit to how compact a mass can become before it can no longer deform. We declare that the TON particle is the limit of compressibility and deformity.

- **Conductivity.** The particle must carry charge to satisfy the requirements for electromagnetism. Therefore, it must exhibit conductive capabilities.

- **Indestructibility and Conservation.** From Newton's laws, we know that all energy and mass must be conserved in all interactions if we are to have a closed universe.

- **Permeability.** The particles must respond to magnetic fields.

- **Permittivity.** The particles must respond to electrical fields. Permittivity becomes a key feature needed by the TON particle to support stellar fusion.

From a scientific analysis of gamma and cosmic rays, we know that the mass of a TON particle is very, *very* small. Based upon our calculations,[9] we have established initial estimates as follows:

Table 1: Estimated TON Particle Parameters

Mass	3.3399×10^{-82} kg
Diameter	8.261×10^{-34} m
Charge	2.8782×10^{-92} Coulombs
Number of TON particles in a proton	5.0078×10^{54}

Looking at these estimated parameters, it is obvious why no one has seen a TON particle. We now move from the unimaginably small to the unimaginably large.

---

[9] See Proof 1 in TON Proof Appendix for these calculations.

The Universal Sphere

We assume our readers are familiar to some extent with the current "Big Bang Theory."[10] Given the observed expansion of the stars and galaxies of our universe, we assume that this idea is at least partially correct—that there was a literal center of the universe billions of years ago, and the universe expanded outward from there.[11]

From this initial starting point, we can imagine the universe at its beginning, when Time = 0 or "Time Zero," as one giant sphere of TON particles,[12] the basic building blocks of everything. Since TON particles are incompressible and indestructible, they can be pressed tightly together, but still retain their basic spherical shape. Because all TON particles in the sphere are touching one another under the influence of an unimaginably strong gravitational force, they cannot move at all and the sphere behaves as one composite TON particle object.[13]

At Time Zero, all energy is potential, not kinetic.

Remember that a single mass object cannot vibrate; it needs another object with mass and charge to oscillate against (page 6). If all mass in the universe is at a single location and touching, then we have no *moving* mass to satisfy Newton's kinetic energy formula:

$$E = \tfrac{1}{2} m v^2 \quad \text{(Eq.2)}$$

If the velocity (v) of the mass (m) equals 0, then $E = 0$. So, at Time Zero, there is no movement, no vibration, and no kinetic energy: all energy in the Universal Sphere is locked up as *potential* energy.

Einstein defined the maximum velocity for any mass as the speed

---

[10] We mean the scientific theory, not the television show of the same name!

[11] Before someone asks, we do not know what surrounded that sphere. All we know is that there was something into which the universe expanded.

[12] The sometimes-postulated notion that all the universe's mass can exist on the head of a pin only demonstrates a lack of comprehension about how much mass it takes to make the universe.

[13] Technically, this Universal Sphere is not solid. There is a "packing" factor that determines how TON particles fit together inside this sphere. We go into this extensively in Proof 1.

of light, (*c*). Therefore, the maximum total kinetic energy for the universe would be equal to the sum of all the mass times the speed of light squared (Eq.1). Since all energy is conserved, at Time Zero we must have an equivalent amount of potential energy locked up within the universal sphere of TON particles, just waiting to be unleashed.

The Two Basic Forces

In Chapter 12, The Big Crunch, we will show how the universe will eventually stop expanding and then contract back to its origin. The stellar fusion process converts all kinetic energy back into potential energy—preserving the conservation of mass and energy. For now, let us explore the source of the forces that can provide the energy needed to drive the universe as a closed system, with mass and energy 100% conserved from the beginning, to expand, then contract, and ultimately cycle forever—unless something changes time and space as we know it. To have a cycle[14], we need to define the repulsive and attractive forces that would cause the universe to expand and contract.

The repulsive force: What repulsive force would force the mass of the universe to move? Luckily, we already know this repulsive force occurs because one of our TON particle properties is that they are conductive and able to hold an electrostatic charge. An electrostatic charge forces any charged mass object to move away from another mass object with charge. If we assume that the electrostatic force is the repulsive force, we can apply Coulomb's Law[15] as:

$$F_R = K_E(q_1 q_2) / R_S^2 \qquad (Eq.3)$$

Where: F_R is the repulsive force,
K_E is Coulomb's electrostatic force constant,
q_1 and q_2 are the electric charge of two TON particles, and
R_S is the distance between the *surfaces* of two charged objects, as shown in Figure 1-1.

---

[14] A cycle implies attraction and repulsion.

[15] Charles Augustin Coulomb, 1736-1806, French Physicist.

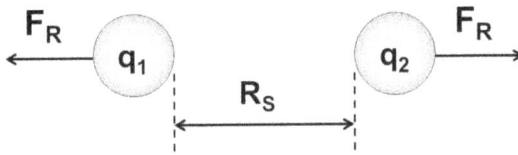

Figure 1-1. Electrostatic Forces Acting on Two Charged Masses

In classic physics, two particles with the same charge will always repel each other. It takes considerable attractive force to push those two charged particles to the point where they touch. So, in free space, like-charged particles would not normally combine without some other forces acting upon them. We can all agree that physics has well documented this basic repulsive force.

The attractive force: Now we add our second, attractive force, F_A, commonly referred to as gravity, which is also well-documented, ubiquitous, and proven to bring any two mass objects together in the absence of any other force acting upon those particles. The *attractive* force of gravity is mathematically very similar to our electrostatic repulsive force. According to Newton's[16] Law of Universal Gravitation:

$$F_A = G(m_1 m_2)/R_C^2 \quad (Eq.4)$$

Where: F_A is the attractive force,
   G is Newton's gravitational constant,
   m_1 and m_2 are the masses of the same two objects we
      used for the repulsive force, and
   R_C is the separation between the *centers* of two objects,
      as Figure 1-2 depicts.

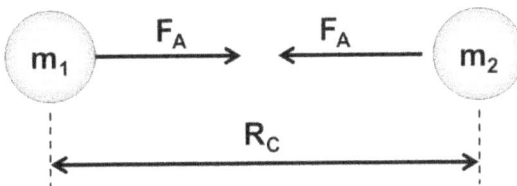

Figure 1-2. Gravitational Attraction of Two Masses

---

[16] Sir Isaac Newton, 1642-1727, British mathematician and physicist.

So, at this point, we have a basic attractive force (gravity) and a basic repulsive force (electrostatic charge) that can control the actions of TON particles throughout the universe.[17]

These two forces can also be shown to work as an efficient spring constant by which two similar mass objects of similar charge can oscillate in free space without any other outside forces acting upon them. Even more importantly, the attractive and repulsive action will occur without any energy loss, which satisfies Newton's conservation laws. This constant vibration is also consistent with scientific observations of photon behavior. In cases where charged particles are moving at some velocity, they create an electric charge flow (current) and consequently a third, magnetic, force as well, which comes into play as we explain in later chapters.

What about the strong and weak nuclear forces?

It turns out that we don't need the so-called strong and weak nuclear forces, because they are artifacts of a theory about atomic physics that, with the introduction of TON Particle Theory, is no longer valid. We will go on to prove this fact, but for now, please keep reading—we are just starting to get to the interesting parts.

Every new idea that goes against conventional thinking may induce some degree of skepticism. So, as you read, consider what Einstein said:

"If at first an idea isn't absurd, there's no hope for it."

— Albert Einstein, from the film *1 Giant Leap*

---

[17] I vaguely remember the first time I saw both formulas. At the time, I instinctively knew that they were somehow connected. Little did I know then that I would be able to eventually combine them into a comprehensive, unified theory. —DAB

If an event emits no detectable energy and creates no sound or light, can it truly be considered a bang?

Chapter 2. The Great Expansion

Before Time Zero and the instantiation of the universal sphere, there was just space. That's pretty much all we can say about it. We really do not know the composition of space. Is it a vacuum? Does it have mass? Is it some new form of energy? We just don't know. The theoretical physics community has established, or at least achieved a degree of consensus, that just before Time Zero there was energy released into the universe by a tear or pin prick in space. As the energy expanded, it began to accrete subatomic particles of mass using an array of "special" particles and energies.

Now, there is a lot of debate as to the nature of the mass objects created in that initial accretion. Some physicists' hold that all mass appeared as a single sphere (or singularity) amidst what Einstein called space-time. Others declare that all mass appeared in the universe as individual particles distributed in a state of maximum universal expansion. One theory has the universe filled with hydrogen and helium gas in addition to a host of interesting particles. Still Other theories have the universe filled with innumerable mass and massless objects. None of these theories have concisely explained how that initial burst of energy created the universe, accreted mass, caused the Great Expansion, or enabled the first stars to appear. TON Particle Theory gives us the tools to answer several of these questions.

Since the stars and galaxies of the universe have been observed as expanding symmetrically from a notional point of origin, we can safely assume that the universe began as a single sphere of TON particles that included all the mass and potential energy of the universe. As we discussed in Chapter 1, we call this object the Universal Sphere (Figure 2-1).

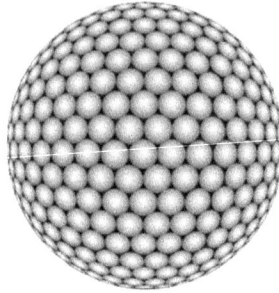

Figure 2-1.
The Universal
Sphere
(not to scale)

The unimaginable gravity of the Universal Sphere had overcome the electrostatic force of the individual TON particles and forced them to touch. TON particles are conductive, so once two TONs touch, all charge is equalized and, while they have the same charge, the touching TONs are considered one particle since the separation distance is $R = 0$ in Coulomb's equation of $F_R = K_E (q_1 q_2)/R_S^2$ (Eq.3). By equalizing all of this charge within the sphere, there can be no repulsive force to provide any TON particle movement, and thus there would be no kinetic energy. (See Chapter 1.)

We speculate that the sudden appearance of the Universal Sphere into space would displace some other form of mass/energy. As Sir Isaac Newton observed, "for every action there is an equal and opposite reaction." So, it does not take a great leap of logic to see how the insertion of the universal sphere into space[18] would displace whatever it is made of and create the expectation that space would push back on the universal sphere with the same force that caused it to be inserted in the first place.

Following Newton's Third Law of Motion, we would see space react by exerting a uniform force along the entire surface of the Universal Sphere, which would initiate a shock wave traveling inwards, that would reach all the way to the center of mass (Figure 2-2). From the

---

[18] Yes, we can endlessly debate the metaphysical implications as to who or what caused the universal sphere to exist and to be inserted into space, but not here.

center[19], the shock wave would rebound[20] and create a minute separation between each TON particle (Figure 2-3).

We only need a small separation[21] (in Eq.3, R_S is greater than zero) between TON particles in the Universal Sphere to instantly reverse the dominant force from gravitational attraction to electrostatic repulsion. So, we return to Coulomb's equation (Eq.3), assume R is greater than zero, and...

Figure 2-2. Shockwave Propagates to Center of Universal Sphere

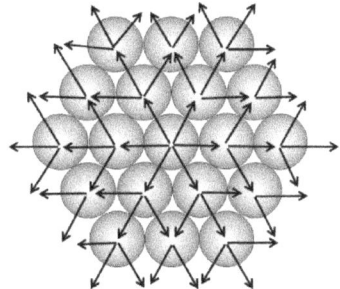

BANG!

Figure 2-3. Shockwave Rebounds Outward from Center of Universal Sphere

Sorry, but we just couldn't resist. Of course, there would not *really* be a bang,[22] but there *would* be an instantaneous expansion force from the surface outwards, as the force on the outermost TON particles causes them to accelerate outward from the sphere, and causes the potential energy of each TON particle to be

---

[19] The initial force along the surface can be very small. As the force wave propagates inward, the energy density increases as the sphere literally focuses all of the force into the center. When all the force reaches the innermost TON particle, the energy involved would be very large.

[20] This is the point where the incompressible and indestructible properties are needed to maintain the physical instance of the TON particles.

[21] This separation is supported by the packing factor of the small TON particle spheres within the Universal Sphere. All those tiny gaps between the TON particle surfaces give them a little wiggle room when the shock wave tries to compress them.

[22] Come to think of it, if it were possible to observe the so-called Big Bang from a safe distance, not only would there be no sound, but there would be nothing to see. The Universal Sphere is the ultimate "dark matter object" and all the TON particles expanding out from it at that moment are "dark matter," because photons (electromagnet radiation, including visible light) have not yet been created.

converted into the kinetic energy of a moving mass object (Figure 2-4).

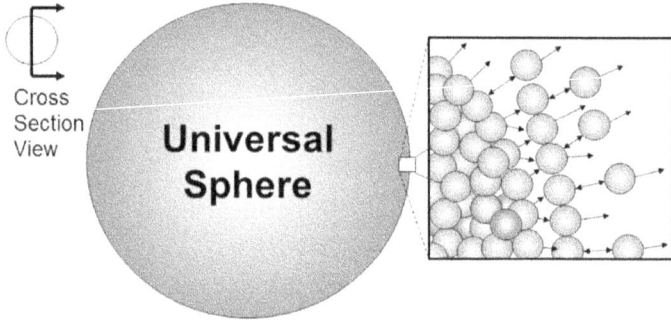

Figure 2-4. Outward Movement of TON Particles After Time Zero

In the following discussion, remember these points:
- We are dealing with incredibly small distances between TON particles.
- We defined the individual TON particle as incompressible, which means that whatever force was exerted inward would rebound with the same force outward without deformation. All energy and mass are conserved.
- The freed TON particles cannot fuse back together because, in this instant, the repulsive force of Eq.2 is much ($\sim 10^4$ times) greater than the attractive force of Eq.4, so everything expands outward from the center of the singularity at an ever-increasing acceleration[23], as Figure 2-5 shows.

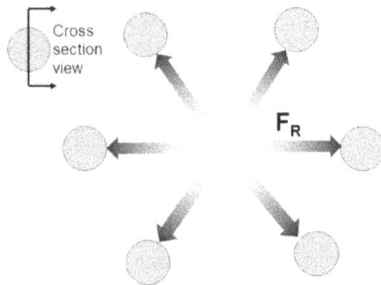

Figure 2-5. Outward Acceleration from Center of Mass

---

[23] Newton's Second Law of Motion.

For those who like to see the math, we get acceleration when the force acting on a mass causes the mass to move, as:

$$F = ma \qquad (Eq.5)$$

Where: F is force
m is mass, and
a is the acceleration caused by the force F, in this case electrostatic repulsion, F_R.

This energy conversion sequence is a very elegant way to account for both why and how the mass of the universe began its outward expansion to overcome the inherently huge gravity caused by the mass of the universe.

As the TON particles of the Universal Sphere separate, they continue to be affected by gravity, and thus attracted to the center of greatest mass. However, each TON particle sees the much stronger electrostatic repulsive force of its immediate neighbors as having the larger force impact. So, each TON particle finds itself with an overall radial force vector (force in a direction) away from the singularity location, as Figure 2-5 previously illustrated. In addition, the initial inertial momentum of each TON particle is deflected in all three directions in a Cartesian coordinate system as the multitude of force vectors builds and falls with TON particle movements (Figure 2-6). At this point in time, the mathematics, though computationally straightfor-

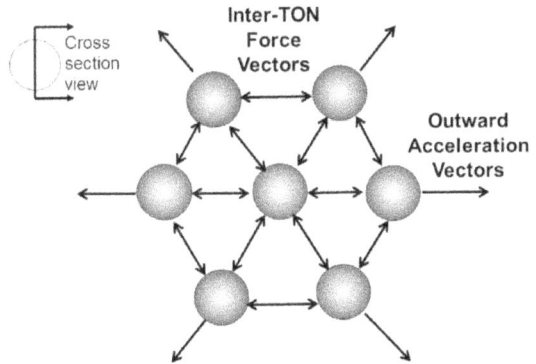

Figure 2-6. Inter-TON Force Vectors

ward, gets really complicated just from the sheer number of TON particles and force vectors involved.

Since we now have moving, charged TON particles, we also have electricity and magnetism affecting all TON particles, as depicted in Figure 2-7.

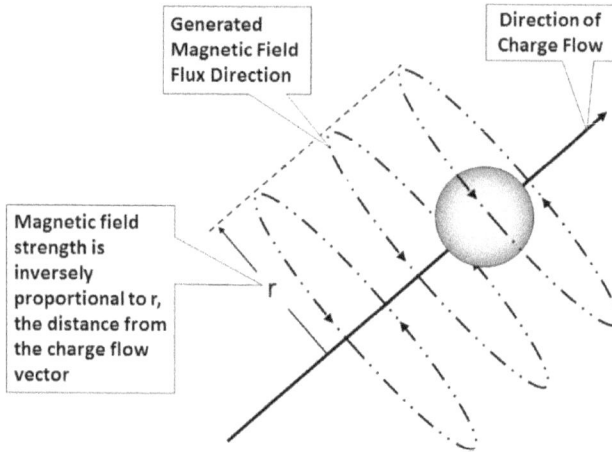

Figure 2-7. Electrodynamic Forces Acting on Each TON Particle

Remember, a moving charge defines an electric current, and an electric current instantaneously generates a magnetic field, as defined in Maxwell's Law.[24] That magnetic field encircles the electric charge movement vector and reduces in strength at longer distances from that vector, just like charge and gravity.[25] The creation of the electromagnetic force adds an angular force component to TON particle interactions.

To summarize, we now have individual TON particles dispersing outward from the original "center of the universe" with increasing acceleration. Three forces affect all interactions between the individual TON particles:

- Electrostatic repulsion (Coulomb)
- Gravitational attraction (Newton)
- Electromagnetism (Maxwell)

Now that we have established how the universe began its expansion, the following chapters will show how these same three forces can harness the massive numbers of TON particles and convert them into all forms of matter at the macro scale that science has either observed

---

[24] James Clerk Maxwell, 1831-1879, British physicist, defined four laws of electromagnetic behavior.

[25] We maintain that the symmetry of these three forces interacting cannot be accidental. They fit hand in glove in an elegance worthy of divine intention. —DAB

or speculated about (including dark matter and plasma). We will begin by showing how a star forms and begins the vital process of fusing TON particles together to:

- reduce kinetic energy,
- restore gravity dominance, and
- return all mass to the universal sphere.

The Big Crunch

At the beginning of this chapter, we noted that there is a theory that all mass "appeared in the universe as individual particles distributed in a state of maximum universal expansion." Since science has observed the expansion of the universe (at least as far as our instruments can detect), it is logical that the universe will indeed reach this point eventually. We will discuss this when we talk about the "Big Crunch" in Chapter 12. There you will see how gravity would bring all mass back to the center of the universe in a reversal of the Great Expansion after the repulsive force is diminished by converting kinetic energy back into potential energy through stellar fusion (see Chapter 3).

*Since a star is really a factory for combining TONs into
TON particle objects, we can redefine it as a
"**S**imple **T**ON particle **A**ccretion **R**eactor."*

Chapter 3. Star Birth

So, what happens next, to get us from a uniform expansion of the universal mass to the universe as we currently know it?

With all the various forces acting upon each TON particle and the electrodynamic effects, all mass eventually starts to collect into small swirls as gravity creates local eddies (small, localized force loops) sufficient to attract masses of TON particles around central locations. One can play with the math, but when we have electromagnetic forces affecting charged particles moving in the same general direction, the particles will begin to rotate around the main outward radial movement vector, which then "herds" the TON particles into giant particle swirls in free space. We have a lot of NASA (National Aeronautics and Space Administration) and ESA (European Space Agency) astronomical photographs of instances throughout the universe to confirm that this swirling takes place. We also have many universe models that predict both how and why it occurs, so again, we are not introducing anything new here. We are just reminding the reader that our theory moves from Time Zero to present day in a very predictable and observed fashion.

As TON particles begin to swirl around their local center of mass due to their electromagnetic properties, we now have TON particles being forced closer together than they would normally be in free space. The swirling charges will also form a very strong magnetic field around the mass of swirling TON particles.

As the swirl concentrates TON particles, the gravitational force created by the overall mass of TON particles draws the innermost TON particles closer and closer together. As the density of TON particles increases at the local center of mass, the gravitational force becomes greater than the individual TON particles' electrostatic repulsive force. Some TON particles will rebound away from this center of gravity if their local electrostatic force finds a weakness in the localized gravitational field. Most, however, will continue towards the center of mass as gravity pulls them inward.

At the same time, the electrostatic force of the TON particles further from the center will tend to push those in the center even closer together, as Figure 3-1 shows.[26]

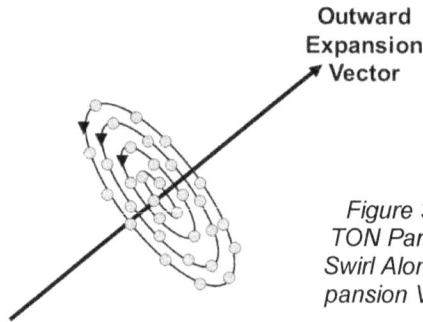

Outward Expansion Vector

Figure 3-1.
TON Particles
Swirl Along Ex-
pansion Vector

We consider the formation of these early TON particle swirls to be the first stage of protostar[27] development. The TON particle swirl at the center has not yet created any TON particle pairs (photons). So, we have no form of emitted, vibrating energy coming from the swirl, and it would be virtually[28] undetectable by any of our current electromagnetic instruments.

Stage 1: TON Particle Expulsion

At the very beginning of the stellar fusion process, two of the innermost TON particles have retained just enough electrostatic repulsive force to expel one or both away from the center before they can

---

[26] To visualize this pushing effect, consider a crowd of people trying to get into a rock concert. As the band begins to play, everybody rushes into the gate. Those in the rear of the crowd push those in front and accelerate their movement into the arena. This is the same thing that happens to TON particles.

[27] The standard definition of a protostar references gas and dust, neither of which have yet to be created at this early point in the universe's evolution. Our definition of a protostar is a collection of TON particles before any photons are created, which is a specific, measurable event.

[28] It might be theoretically possible to detect such a swirl/protostar if our view of it happened to be backlit by a light source. It might, for example, show up as a tiny black dot against the background of a glowing gas cloud or nebula.

touch, as Figure 3-2 shows.[29]

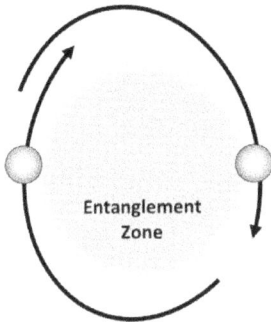

Figure 3-2.
TON Particles Ap-
proach
Entanglement

A good example to illustrate this stage is the repulsion of two bar magnets when placed north pole to north pole. No matter how hard you push the two together, the magnetic force will always push them apart. Think of the effect as two wet marbles being squeezed together with wet fingers. You reach a point where both marbles will pop away from the fingers from the pressure. This is the first sign that the protostar is about to initiate stellar fusion.

As the combination of forces increase in the center of mass, a pair of TON particles will enter into a gravitational and electrostatic entanglement[30], which locks them together, although they are still not touching. The "entanglement zone" depicted in Figure 3-3 is an area where TON particles are forced so closely together that they become affected by gravitational and electrostatic near field effects. In this near field area, the two objects behave as a single object, even though

---

[29] Note, these TON particles will be swirling with an angular velocity. The interaction with the repulsive force will convert some of that angular momentum into linear acceleration.

[30] We define an "entanglement" as two objects of similar mass and charge in such close proximity that the electrostatic and gravitational force vectors form an effective "spring" between them. This spring enables the two TON particle objects to vibrate against each other in free space, but it initially firmly binds the two TON particle objects together as a rotating pair around a mutual center of mass. The result of entanglement is the creation of a photon.

they are not physically touching[31]. Again, remember that we are talking about incredibly small distances here. After the entanglement occurs, the protostar has created its first *unit photon*, composed of two
single TON particles. This is the smallest (i.e., least massive) photon
possible.[32]

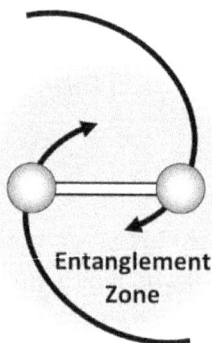

Figure 3-3.
Two TON Parti-
cles Entangled to
Form a Unit Pho-
ton

Obviously, the new photon (entangled TON particle pair) would
have the mass of two TON particles. However, the functional properties of this photon are significantly different than those of individual
TON particles and free photons. Specifically, as the TON particles are
forced very close together (approximately 10^{-15} meters), the two TON
particles will lock in rotation around each other by combining gravity
and the electrostatic repulsive force between them (Figure 3-3). (See
Proof 2 for the mathematics involved.) At this point, the surrounding
stellar gravity is insufficient to compress the two objects further, and
the electrostatic shear forces and swirling of the proto-star mass will
force the two TON particles to spin around their localized center of
mass.

The two rotating TON particles form an electromagnetic toroid

---

[31] Experiments have confirmed that, when small charged objects are forced together,
the electric field envelops both, such that they appear as a single charge source from
far field distances. It is logical to assume that the gravitational field will behave similarly at these distances.

[32] Once we establish the final mass of a TON particle, we can base all the other
property calculations that involve photons on this value. In the rest of this book, we
use the unit photon as the basis for calculating photon interaction. (See "The Calorie
and the Photon" on page 78.)

(further explained in Proofs 2 and 4) and form an electromagnetic field around the new photon. This electromagnetic field is the key to the photon's survival. The electromagnetic force cancels most of the electrostatic effects of the surrounding TON particles. This increased immunity from these electrostatic forces permit the photon to avoid some of the further fusion effects, and enable it to eventually escape out into the outer edges of the star because its enclosed volume makes it less dense than the surrounding collection of individual TON particles.

Some of these newly formed photons begin to escape the stellar core and form the stellar photosphere.[33] While in the photosphere, the photon retains its compressed state, which we describe in more detail when we talk about atomic photon absorption in Chapter 7. The photon will remain in this state until it is expelled into free space and escapes the star's gravity. Since the compressed TON particles are rotating around, rather than vibrating against, each other, we will not be able to detect its vibration frequency until then. As noted in "Today's Geometry Lesson," due to the difference in density, the larger mass photons will migrate to the star's surface. From there, they can escape the star's gravity into free space and begin to vibrate in the visible and infrared regions of the spectrum. This phenomenon also explains sunspots as an upwelling of TON particle objects and less massive photons from the deeper levels of a star. When they reach the star surface, they will appear as dark spots (sunspots). When emitted into space, the single TON particles do not vibrate at all, and the less massive photons vibrate at a frequency higher than "visible light." Likewise, apparent "holes" in a star's corona are really emissions of these "invisible" TON particle objects.

Today's Geometry Lesson

When the photon forms, the radius of the photon is determined by the mass and charge of the two TON particle objects inside the photon. As those photon particle objects get larger, the radius increases and the encapsulated volume of space increases at a faster rate than the mass increase. Therefore, the most massive photons will always be less dense than less massive photons and thus will migrate to the surface of the star.

---

[33] It is called a photosphere because of the collection of photons that surround a star.

When the unit photon escapes a star's gravity, it will vibrate as shown in Figure 3-4 at an extremely high frequency. Our initial calculations in Proof 2 put these frequencies in the 10^{30} to 10^{40} Hertz range. In fact, photons both *vibrate and rotate* around a central axis, which creates different properties (polarization), but we will defer the rotational aspects to later chapters.

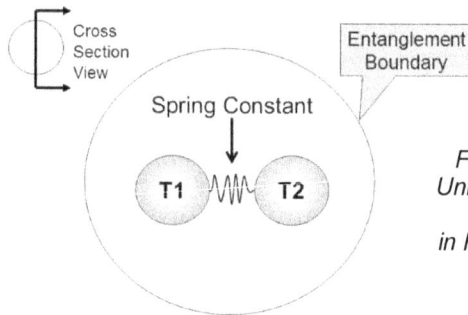

Figure 3-4.
Unit Photon Vi-
brating
in Free Space

Why Photons Vibrate

Oscillation theory in physics supports TON Particle Theory's assertion that photons are composed of two entangled, vibrating TON particle objects. We know that two objects of some equal mass could vibrate if they were placed under an attractive and a repulsive force in free space. We also know from classic physics that the more mass we have, the slower the two particles will vibrate, as the energy needed to accelerate and decelerate the mass increases. If we review oscillation theory for objects, we find that the frequency of oscillation is equal to the inverse of a time quantity, which in turn is a function of a measure of elasticity, like a spring constant, and the mass, as:

$$f = 1/T \qquad \text{Eq.6}$$

Where: f is the frequency of oscillation, and
T is the period of oscillation, and

$$T = 2\pi/(K/m)^{0.5} \qquad \text{Eq.7}$$

Where: K is a spring constant or some similar elasticity constant, and
m is the mass of each vibrating object.

So, if we can take two objects of equal mass and put them under a stable attractive and repulsive force situation, we can get them to vibrate at a fixed frequency until an external force or mass disrupts them.

To summarize this section, stars are the factories of the universe. Stars create all of the more complex objects in the universe from the first unit photon to the first large dense mass object (LDMO), a.k.a. "black hole," and everything in between. In Stage 2 stellar fusion, the protostar took two TON particles and manufactured something new— a photon. Therefore, in TON Particle Theory, we now define it as a star, even though *true* stellar fusion has not yet occurred.

Now we will examine how a new star begins true TON particle fusion.

Stage 3: TON Particle Fusion

When two spherical, charged objects (TON particles) begin to approach each other within approximately five to ten diameters apart, an interesting phenomenon occurs[34]. The electrostatic force vectors between the two objects begin to cancel, which decreases the gravitational force required to make them touch. In the third and final stellar fusion stage, gravity forces the two TON particles of a unit photon together until the distance separating them goes to zero, as Figure 3-5 shows. At this point, the star creates a different type of gravitational and electrodynamic link, a true particle fusion.

---

[34] When electric fields encounter a conductor, the field is absorbed depending upon the permittivity of the material. Since a TON particle is a pure conductor, the field is completely absorbed, thus canceling some of the repulsive force between the two objects.

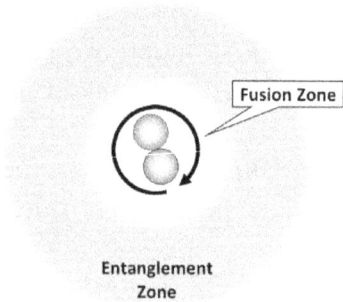

Figure 3-5.
Star forces compress a
Unit Photon, fusing its
TON Particles into a
TON Particle Object.

Entanglement
Zone

As we have said earlier (page 18), each TON particle is conductive, has ideal permittivity, and is permeable. Once they physically touch, their combined charge is distributed around the resulting surface, the repulsive force between the objects goes to zero, and gravity holds them together with their maximum attractive force. Since TON particles are incompressible and indestructible, each TON particle retains its individual spherical state, but the two fused TON particles can now be treated as a single TON particle object with the mass of two TON particles.

Unlike a photon of the same mass, the new TON particle object does not vibrate when emitted into free space. The two TON particles are firmly locked together, but still spinning around each other. The overall charge of the object will be less than that of the photon, because each of the two objects absorb part of the electrostatic force by conducting part of the electrostatic field emanating from the surface of the other object. That force is blocked by the physical orientation of the two touching objects at the point of intersection, as a "shadow area" as shown in figure 3-6.

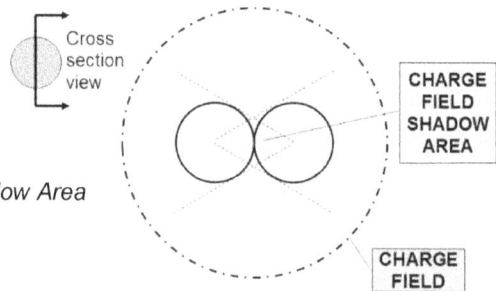

Figure 3-6. Charge Field Shadow Area

Part of the kinetic energy represented by the electrostatic repulsive force of the original two TON particles is now stored as potential energy, which is released only when the two objects are forcibly separated during nuclear fission. The angular momentum of the spinning object retains the rest of the kinetic energy.

Nuclear fusion converts kinetic energy into potential energy.[35]

At this point, it's important to review the activities occurring at the center of the star. The accumulation of TON particles has created a localized center of stellar mass sufficient to overcome the electrostatic repulsive force between two TON particles. The fusion process captures some of the energy related to that electrostatic force and stores it in the newly formed, fused TON particle object as potential energy. Therefore, it takes a significant external force to split the two TON particles apart, as they now experience the maximum amount of gravitation attraction possible between the two objects. It is this stored potential energy that is released during nuclear fission.

Nuclear fission releases potential energy and converts it back into kinetic energy.

We can see a representation of this stored force by reviewing Newton's gravitational equation (Eq.4) from Chapter 1:

$$F_A = G(m_1 m_2)/R_C^2 \qquad \text{(Eq.4)}$$

Where: F_A is the attractive force,
G is Newton's gravitational constant,
m_1 and m_2 are the masses of two objects, and
R_C is the separation between the centers of two objects

When the separation distance (R_C) between the centers of mass of two TON particles is equal to the diameter of a TON particle, we can use this formula to calculate the maximum gravitational attraction

---

[35] This conclusion explains and confirms the results of nuclear fusion experiments conducted since the 1960s, where it has been proven that nuclear fusion does not release excess energy. Therefore, the dream of clean, nuclear fusion power generation will always remain a dream unfulfilled.

that can occur between those two TON particles. As we discussed earlier, as soon as these two TONs separate at any distance greater than zero (greater than the diameter of one TON, in this case), the electrostatic force is manifested and is 10^4 times greater than the existing gravitational force. The electrostatic force vector then rapidly accelerates the TON particles apart, converting the stored potential energy into kinetic energy. So, in nuclear fission, that force balance change is what powers the disintegration of the atom and releases the stored energy.

Stellar Fusion: Stage 1, Round 2

As Figure 3-7 illustrates, along with TON particles from the Universal Sphere, the stellar core also now contains the first unit photons and the first fused TON particle objects, each with the combined mass and charge of their original two TON particles (a 2-TON particle object). The star is now ready to begin the second round of stellar fusion.

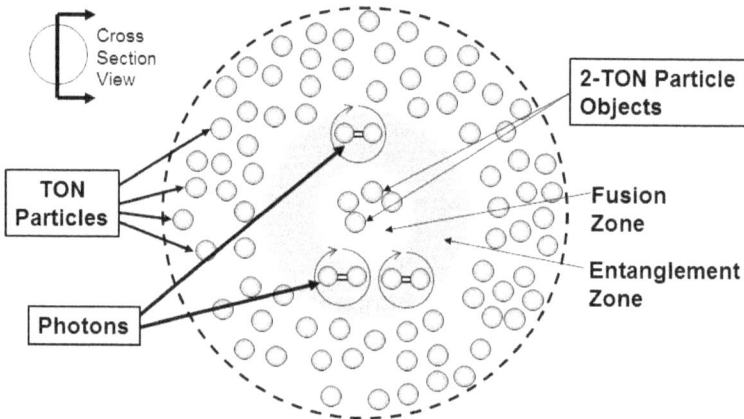

Figure 3-7. Fusion Round 1 Complete

The gravitational force created by the overall mass of TON particles in the stellar core draws the innermost TON particle objects closer and closer together, initiating the next stellar fusion phase. Once again, the electrostatic charge is still just strong enough to prevent TON particle objects from touching.

Stellar Fusion: Stage 2, Round 2

As gravity increases, two 2-TON particle objects enter entanglement and become a photon with a mass equivalent of four TON particles (Figure 3-8, EZ-2). The star has now created its second photon type. The larger mass and charge of the new photon creates a vibration frequency that will be lower than the vibration frequency of the previously created unit photons.

Stellar Fusion: Stage 3, Round 2

Next, gravity increases to the point where the 2-TON particle objects in a 4-TON photon are compressed and fused together (Figure 3-8, FZ-2) to form a 4-TON particle object having the mass and charge of its four TONs. This denser TON particle object sinks deeper into the center of the star as its mass is drawn in by gravity. This completes Round 2.

Stellar Fusion, Round 3

Round 3 takes these 4-TON particle objects, and they go through the same three stages as in Rounds 1 and 2, resulting first in photons with mass and charge equivalent to eight TON particles, which are then fused into 8-TON particle objects (Figure 3-8, EZ-3 and FZ-3). This three-stage fusion process will repeat as long as the star has enough gravity to fuse larger and larger photons and TON particle objects. [36]

Primary Fusion Sequence

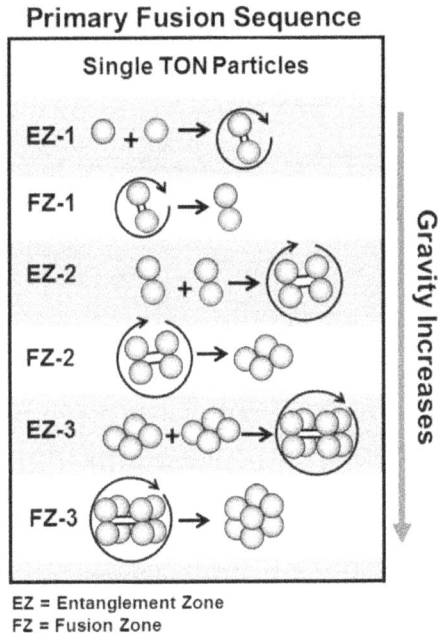

EZ = Entanglement Zone
FZ = Fusion Zone

Figure 3-8. Primary Fusion Sequence

---

[36] *In Proof 4 or 5, we go into more detail about how these individual TON particle spheres collect geometrically. Once we fuse approximately thirty-two TON particles, the resulting shape will remain roughly spherical throughout the remainder of the star fusion process.*

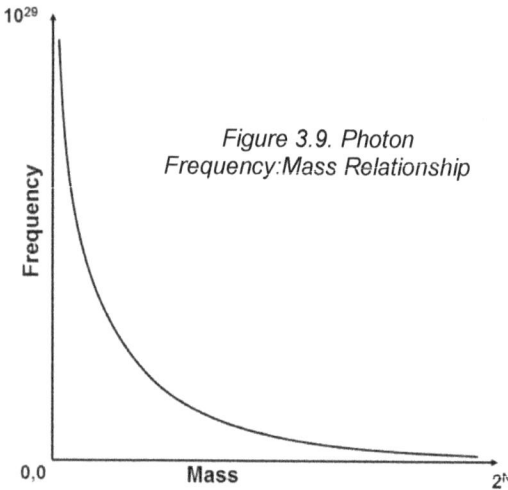

Figure 3.9. Photon Frequency:Mass Relationship

The geometrical progression of photon development creates a series of photon frequencies that get lower as the mass increases. The lower frequencies are caused by TON particle objects and their greater moments of inertia,[37] as Figure 3-9 illustrates. Eventually, the frequency will reach what we call ultraviolet light, then visible light, and then continue into the infrared and millimeter-wave frequencies, as Table 2 shows.

Table 2: Photon Mass Estimates

Type	Wavelength	Frequency	Mass at Specific Frequency
Gamma	2.99×10^{-11} m	10^{18} to 10^{40}	2.5×10^{-39} Kg @ 10^{20} Hz
X-Ray	2.99×10^{-9} m	10^{17} to 10^{20}	2.5×10^{-35} Kg @ 10^{18} Hz
UV	2.99×10^{-7} m	10^{15} to 10^{17}	2.53×10^{-31} Kg @ 10^{16} Hz
Visible	5.98×10^{-6} m	10^{14} to 10^{15}	1×10^{-28} Kg @ 5×10^{14} Hz
Infrared	2.99×10^{-4} m	10^{12} to 10^{14}	2.5×10^{-25} Kg @ 10^{13} Hz

Zones of Stellar Fusion

The idea of TON particle fusion with various numbers of fused TON particle objects resulted from a look at the probable interaction between different fusion zones created within the star. Think of a star as an onion, where TON particle objects of the same mass arrange themselves into bands around the stellar center of mass as Figures 3-7 and 3-8 previously illustrated. There are no hard lines between these zones, and a certain amount of mixing almost certainly occurs.

Boundary Level Entanglements

So far, we have only dealt with the entanglement of TON particle objects of like mass and charge to create photons (Stage 2). It is im-

---

[37] Moment of inertia is equal to the Mass times Velocity of the TON particle objects.

portant to note here that there can also be boundary-level entangle-
ments between the primary sequence of evenly numbered TON particle objects to create the odd numbered photons shown in Figure 3-10.

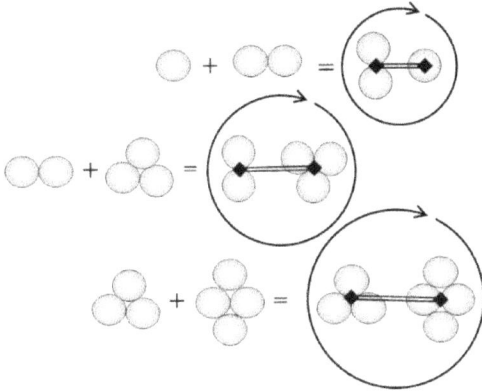

Figure 3-10.
TON Particle Entangle-
ment Across Bounda-
ries to Make Odd Num-
bered Photons

At the boundary zones, it is possible for a 2-TON particle object
to entangle with a TON particle if it goes outward or with a 3-TON
particle object if it moves inward. This would result in a photon with
the mass of three or five TON particles, respectively. In the third primary fusion stage, gravity can compress such an odd-number photon
and fuse it into a three- or five-TON particle object. Such boundary
fusion events can create the full range of TON particle objects, which
we explore in more detail in Chapter 7.

In summary, the primary and boundary entanglement/fusion
events will continue as long as the star has sufficient mass to continue
to make larger TON particle objects. If massive enough, the star will
eventually make TON particle objects with the mass and charge of a
proton, which will eventually combine with photons to become the
first known atomic element, hydrogen.

If you have a particle that cannot be located in space, has charge and mass but appears everywhere at once, then maybe you are not dealing with a single object.

Chapter 4. The Photon Shell

Now let's look at our currently accepted subatomic particles, beginning with the electron. Classic physics defines an electron[38] as a negatively charged particle. What if the electron, as defined in classic physics, is not one particle but a thin spherical "shell" of many objects, specifically photons?

Let's look at how an object the mass of the electron particle could form inside a star. In Chapter 3, we introduced the notion that a star begins to fuse individual TON particle objects into larger and larger TON particle objects. In our star model, it is reasonable to expect the less massive TON particle objects and photons to accumulate in the star's outer layers, as Figure 4-1 shows. Classic physics calls this layer the photosphere because of the photons observed and measured there.

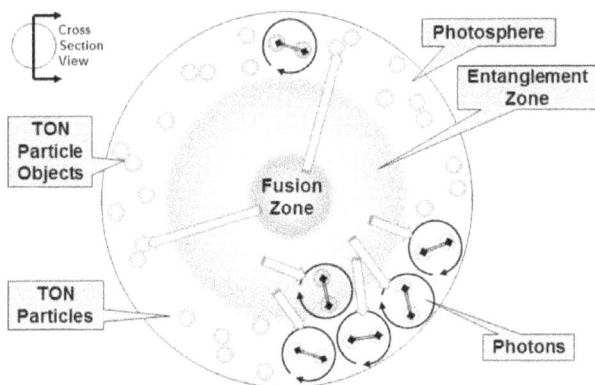

Figure 4-1. TONs, Photons, and TON Particle Objects in the Photosphere

Think about having layer after layer of different-mass photons with fused and singular TON particle objects flowing up and down the

---

[38] Sir J. J. Thompson in 1895.

different layers between the fusion zones and the surface of the star. Now think about a pot of boiling pasta water. Have you ever watched the oil and starch particles form foam on the water surface? That effect is similar to what we would see if a large number of photons rose up together from the depths of the star. As they rose towards the star surface, the reduced gravity, electromagnetic, and electrostatic forces would allow the photons to form a self-supporting, bubble-like shell, as shown in Figure 4-2.

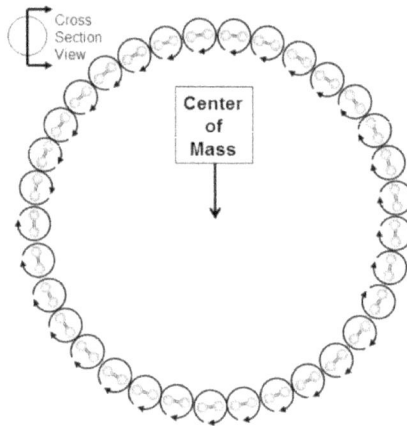

Figure4-2.
Photon Shell
Equivalent to an
Electron

In this instance, we would have a small collection of photons pulled into little hollow spheres by small areas of local gravity, electromagnetic, and electrostatic force (Proof 3). Then we would have tangential forces (magnetic fields created by the rotating TON particle objects inside each photon) along the shell surface holding each photon in an orientation tangential to the surface. All the photons on the surface of the shell would have charge and would be observed at the macro scale as a single particle. Sounds a lot like the classic description of an electron, doesn't it? This leads us to:

"Electrons" are shells of photons.

The electrostatic force created by TON particle objects within each captured photon would generate enough repulsive force to keep the shell from collapsing, as Figure 4-3 shows. Based upon the force balance, TON particle objects within each photon would be constrained to rotate parallel to the shell surface and create a magnetic field perpendicular to the shell surface. Since the lines of electrostatic force would be outward, the photons, regardless of mass, would rotate in the same direction, thus creating the magnetic field around each photon and around the entire photon shell. The summation of the forces involved would then support the shell structure as a stable entity, as Proofs 3, 4, and 5 explain in more mathematical terms.

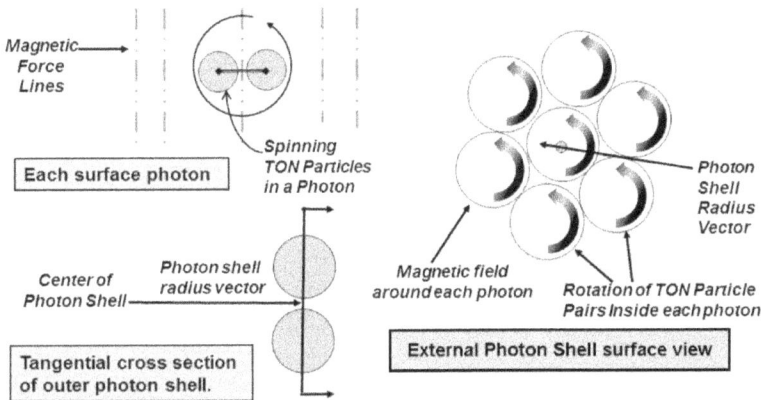

Figure 4-3. Photon Structure on the Surface of a Photon Shell

The same forces that have faithfully served to bring us to the point of stellar fusion can also be used to form both stable photon shells and fused TON particle objects. We can show that the classic electron is probably not a single particle at all, but a stable photon shell. (We will go into this concept in more detail later.

Enrico Fermi discovered small objects inside a proton when he first collided them together. Since then, identifying these objects has become a major pursuit in particle physics. However, there is a simpler explanation for what has been measured and observed.

Chapter 5. The Proton

The classic proton is essentially a TON particle object with the mass attributed to the measured proton. In Chapter 3, we explained how the stellar fusion process would continue to make larger TON particle objects. If the star has sufficient mass, it will produce TON particle objects the size of the classic proton. (Figure 5-1).

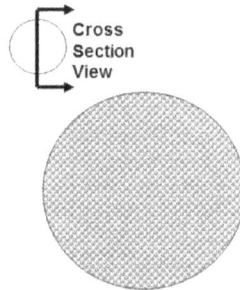

Cross Section View

Figure 5-1.
Proton Composed of
Approximately 10^{54}
Fused TON Particles

Current physics theory has introduced a wide range of sub-sub-atomic particles based upon measurements and analyses of the debris resulting from smashing two proton streams together in a super collider. TON Particle Theory better explains the measurements and data but takes issue with the bevy of newly-named, subatomic (imaginary) objects.[39]

When you do ballistics analysis on two spheres of spheres colliding, the results are very predictable (Figure 5-2). Using TON Particle Theory, we can fully predict what artifacts would be seen during a proton-on-proton collision.

---

[39] For a reasonably current list, see https://en.wikipedia.org/wiki/Timeline_of_particle_discoveries

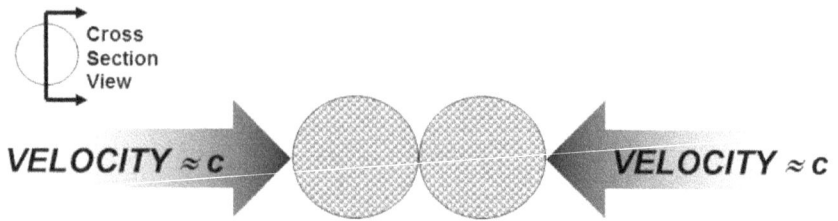

Figure 5-2. Two protons, moving at nearly the speed of light, collide.

First, each proton would be like a sphere of brass balls (BBs)—a lot of BBs (representing approximately 10^{54} TON particles). If you get a perfect hit, the resulting force vector will penetrate both protons and cause a cone of TON particle objects to eject out the back of each proton (Figure 5-3). The amount of mass ejected will be somewhere between 25% and 30% of each proton.

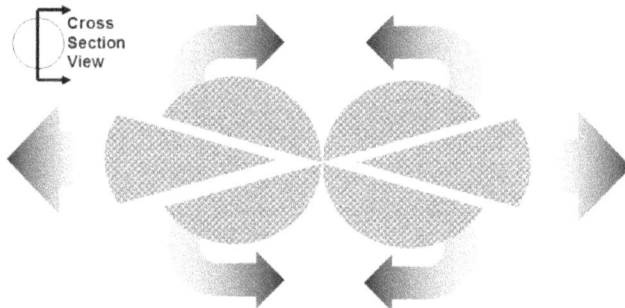

Figure 5-3. A cone of TON particle objects ejects out the back of each proton.
(Time 0)

The remaining 70-75% mass of both protons begins to fracture and fold over the collision zone, creating fractured, curved fragments of TON particle objects in addition to the release of many smaller TON particle objects. The curved plates, like fragments of a cracked egg shell, will be somewhat similar, but will vary in size and mass, depending on the fracture pattern caused by the specific way the individual TON particles are arranged in each proton and fracture object (Figure 5-4)[40].

---

[40] The vertical lines represent the time steps used to describe the specific impacts of each numbered proton fragment.

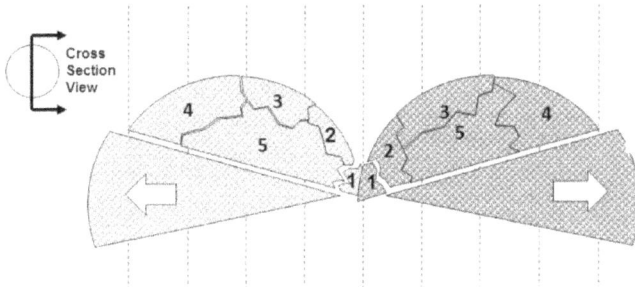

Figure 5-4.
Proton Collision
(Time +1)

At Time +2 (Figure 5-5) fragments labeled 1 have rebounded from their collision point. Fragments labeled 2 are now colliding. However, like a train wreck, momentum forces fragments 3 and 5 to also collide with fragments 2. The secondary collisions may cause additional fracturing and TON particle object dispersal from each of the fragments. Since fragments 4 are still loosely connected to fragments 5, the collisions between fragments 5 and 2 will also affect fragments 4.

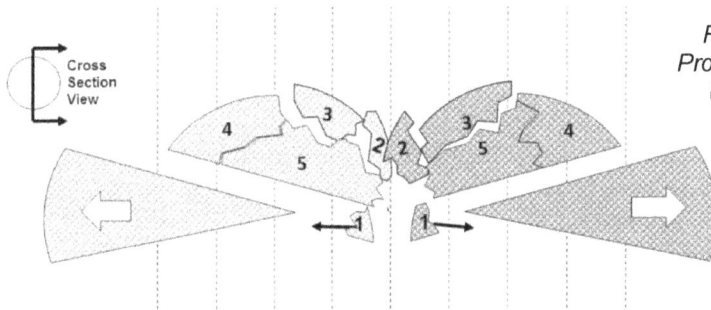

Figure 5-5.
Proton Collision
(Time +2)

At Time +3 (Figure 5-6) we have the secondary collision between fragments labeled 3. Meanwhile, fragments labeled 2 have rebounded

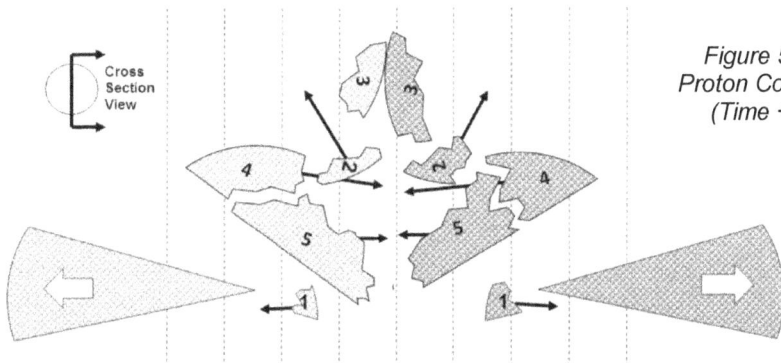

Figure 5-6.
Proton Collision
(Time +3)

and are trying to escape the collision zone. Fragments 4 and 5, now fully separated, continue to advance towards the collision zone. There may be additional interactions between fragments 4 and 5 as they tumble. Note, fragments 4 may also collide with fragments 2, depending upon their exit trajectories.

Now, at Time +4 (Figure 5-7) fragments labeled 5 collide, while fragments 1 through 3 should be free of the collision zone. Fragments 4 should retain sufficient momentum to continue towards the collision zone, but should be now separated from all the other major fragments. At this point in time, the major fragments will continue to maintain what is left of their physical integrity.

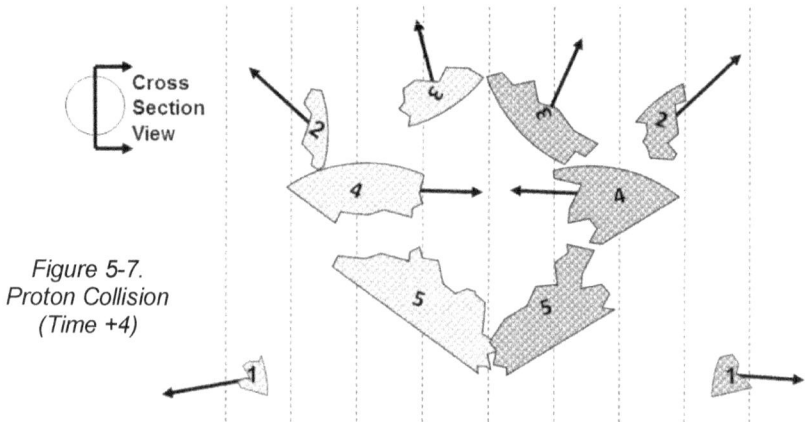

Cross
Section
View

Figure 5-7.
Proton Collision
(Time +4)

Note, there will be a lot of smaller fragments that will fly away from the collision zone at high velocity, and these will leave straight-line velocity trails on the electromagnetic detectors and recording instrumentation. The larger proton fragments will have much less linear velocity and will have time inside the measurement area to interact with the magnetic field caused by the charged protons themselves. The movement of these TON particle plates will leave a curlicue pattern on the collider instruments as the induced magnetic forces introduce additional angular momentum to each fragment.[41] Also, note that some of these fragments will tumble and rotate around their own

---

[41] For spectacular images of these patterns, search the Internet for "subatomic particle tracks."

local center of mass, which matches the pictorial representation of the experimental data produced by super colliders.

The x-ray plates from the early proton-on-proton collision experiments conducted by Nobel Laureate Enrico Fermi clearly exhibit these exact artifacts, though at very low spatial resolutions. When you compare that data to the recent data collected at the Large Hadron Collider at CERN, the European Organization for Nuclear Research, you still see the same type of artifacts only in higher definition. We maintain that, instead of being separate subatomic particles with individual characteristics (Fermions, quarks, leptons, bosons, etc.), these artifacts are merely leftover, random clumps of proton fragments. In short, they are proton (TON particle object) shrapnel.

Consider what happens if you smash two snowballs together at high velocity. They both will shatter into clumps of snowflakes. The clumps and snowflakes may appear different, and every snowflake may appear to be unique, but they are all composed of the same basic ice crystal material.

We have offered a very simplified explanation of what happens when two protons collide. However, inside a super collider, we are dealing with a stream of protons coming from both directions. So, the ejected mass from the first proton hit will impact the trailing protons in both directions, changing their velocity, mass, and charge. Basically, the result is a chaos of ricocheting TON particle objects and a contaminated measurement area. The multiple collisions, ricochets, velocities, electromagnetic interferences, angles of impact, nearly infinite sizes of TON particle objects, etc. — none of which can be directly observed or accurately measured — makes any analysis of the measured data questionable.

Note: no *photons* are emitted in the collision zone, because the proton has only fused TON particles within it. If we could watch the collision with a video camera, no *vibrating* energy particles would be seen emitted from the collision point, nor would our spectrometer show any spectral lines emitted by the protons during the collisions. However, we *could* have photons emitted by atoms in the *containment vessel* that are struck by the various TON particle objects expelled at the collision point. These photons would register to detection equipment as the full electromagnetic spectrum of gamma rays, X-rays, visible light, infrared light, etc.

Classic theory portrays a proton as having a positive charge and an electron as having a negative charge. In reality, both have the same polarity of charge. Remember, electrostatic force is always repulsive, therefore, the notion of positive and negative charge is incorrect. We can only have different amplitudes of electrostatic repulsive force.

Is hydrogen really the starting point of stellar fusion?

Chapter 6. Hydrogen, the First Atom

In the classic atomic model, a proton and an electron particle are combined to create a hydrogen atom. Having an atomic number of 1, with one proton and one electron, the hydrogen atom is listed as the first atomic element in the classic periodic table.[42] The proton is assumed to be positively charged, while the electron particle is presumed to be negatively charged.

In TON Particle Theory, we replace the notional orbiting electron particle with a shell of captured photons and call it the photon shell. It is electrostatic force and gravity that balance the mass and charge of the photon shell around the mass and charge of the proton to form a stable, spherical structure. If the proton and the photon shell had different polarities of charge, as classic theory states, this balancing could not happen. This physical symmetry is the only way to form a stable photon shell around the proton (Figure 6-1).

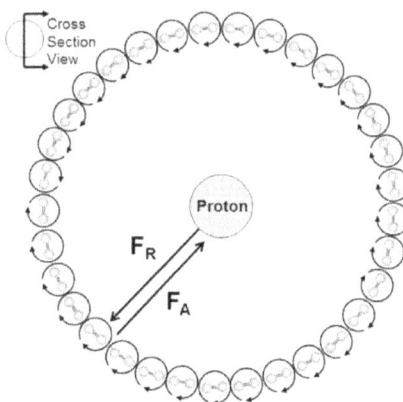

Figure 6-1.
A Hydrogen Atom
Using the Photon
Shell Solution

To begin producing hydrogen, a star must first fuse a TON particle object the size of a proton. This process requires *a lot* of fusion events—assuming a geometrical progression (Chapter 3), we estimate

---

[42] The classic star model supposedly begins stellar fusion using the proton of the hydrogen atom to create the nucleus of the helium atom. In Chapter 7, we will show why this fusion sequence cannot be true.

10^{27} fusion layers must be created before the star begins to fuse TON particle objects the size of a proton (5.0079 x 10^{54} TON particles).

After the star begins to accumulate proton-sized objects, it will enter into the three stages of fusion (Chapter 3) to form the next larger-sized object.[43] As we have already described, in the first fusion stage, the proton-sized TON particle objects will initially fail to entangle and will be ejected out of the fusion zone and into the photosphere. It is this journey into the outer regions of the star that creates the hydrogen atom.

As the proton goes through each layer of photons, it will begin to pull photons (or photon shells) into its localized gravitational field. Like coating an apple with caramel, the photons will envelop the proton but not stick to its surface. As the captured photons encircle the proton, the added mass of the photon shell would firmly bind the shell around the proton due to gravity, but the electrostatic force between the proton and each photon would prevent the photons from being fused to the proton surface, as depicted in Figure 6-1 above. Such a physical arrangement would easily satisfy the theoretical need for the mass and charge of an electron to surround the proton to form a stable hydrogen atom.

The photon shell model easily explains why Nobel Laureate Werner Karl Heisenberg could not measure the position of a single electron particle around an atom. In his frustration, he developed his "uncertainty principle." Had he instead concluded that the electron is not a single particle at all, as his measurements indicated, but a shell of smaller objects, he would have been in a position to discover from his data that he was instead observing and measuring the positions of captured photons around the atom. His data fully supports the photon shell model, and his measurements are within the range resolution of the size of a captured photon in the photon shell.

---

[43] Depending upon the star that first produced them, we might even find some very interesting variations on sub-proton sized TON particle objects that could form stable atomic structures similar to hydrogen but with less mass. Since we are not sure if such sub-hydrogen elements exist, they become a potential new atomic object for science to find.

It is also important to note here that the next stage of stellar fusion would create a photon from two proton sized TON particle objects. These entangled protons are the first step in creating the next level of atomic nuclei. If freed from the star, these massive photons would theoretically vibrate at a frequency of approximately 40 kHz.

Hydrogen and the Four States of Matter

Now that we have established how a photon shell could replace the classic notion of a single electron orbiting a proton in an atom, we can better explain the photon interchange between atoms.

The photon exchange process dominates the universe around us. All active physical, chemical, and biological processes require photon exchange. Photons are absorbed and emitted from every atomic element at specific frequencies; in spectroscopy, this phenomenon is observed and measured as spectral lines. Using the photon shell model, we can easily demonstrate how photons of various masses and charges, vibrating at specific frequencies representing the entire electromagnetic spectrum (visible light, infrared, ultraviolet, Gamma rays, X-rays, cosmic rays, etc.), would be readily available in every atom.

The structure of the photon shell enables a simple process to have new photons absorbed by an atom, as discussed later in Chapter 8, thus expanding the photon shell. The same structure supports the emission of photons, which causes the shell to contract slightly. The degree of expansion/contraction occurs in very specific amounts (i.e., quanta) based upon the specific mass and charge of the photons involved. Scientists have measured and observed the photon shell expansion and contraction, and the effects are well-documented in scientific literature.[44]

---

[44] The term used is "electron shell excitation" in classic physics.

In Chapter 1 we described a relationship showing how the electrostatic forces could balance with the gravitational force at localized (atomic) levels (Figure 6-2). Throughout the photon exchange process, the atom will always try to maintain the balance of these forces at the lowest mass/charge state. This last point is key, because it is this balancing act that determines the physical state of each atom. Capture and emission of photons can only occur if those photons are part of the photon

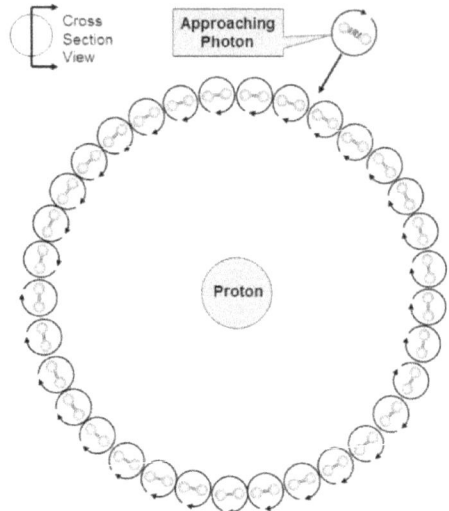

Figure 6-2. Photon Approaches Photon Shell

shell around the atomic nucleus. Attempts to form and destroy photons dynamically in the laboratory have required current scientific theories to introduce a fair number of imaginary particles and forces. TON Particle Theory now better explains the observed phenomena and provides the structure and processes by which atoms can change states among solid, liquid and gas by adding or subtracting photons in the photon shell.

We have already established that a photon that vibrates in the visible light spectrum is a very small subset of all possible photon objects created by stars. The only remaining issue is to understand how the photon shell absorbs or emits *specific* photons.

Using the hydrogen atom as our basic model, we can exchange photons of specific wavelengths and cause the hydrogen atom's photon shell to expand or contract by specific amounts. The overall photon exchange effect is well-known and documented, as science uses this method to stimulate different materials to generate specific wavelengths during lasing. We now explain how and why it works.

Photon Absorption (Capture)

As a photon approaches a hydrogen atom, we get some interesting effects. First, the gravity of the atom interacts with the two TON particle objects vibrating inside the approaching photon, as in Figure 6-2.

In Chapter 3, we described how stellar gravity compresses the spring balance between the two TON particle objects inside the photon, to the point where vibration ceases. Instead, when compressed in this manner, the two TON particle objects will begin to rotate around each other within the photon entanglement boundary.[45] This compression reduces the overall entanglement diameter of the photon.[46]

The localized gravity of an atom has the same effect on an approaching photon. Compressing the spring constant reduces the entanglement diameter of the photon and enables it to more easily fit among the other captured photons in the photon shell, as Figure 6-3 shows. In addition, while gravity is affecting this compression, the electrostatic charge difference between the photon and the atom slows the photon's velocity so that it smoothly intersects the photon shell with only minor physical perturbation to the force balance in the photon shell.

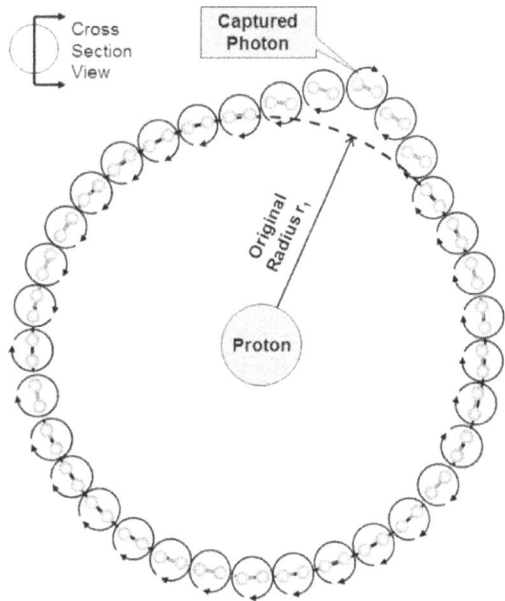

Figure 6-3. Photon Captured by Photon Shell

---

[45] In effect, the photon is converting its vibrational energy into angular momentum to maintain 100% conservation of energy.

[46] We are referencing the photon spin diameter at Bohr's radius. Movement of the photon closer to the atomic nucleus would result in even smaller diameters of spin.

When absorbed into the photon shell, the newly captured photon causes its adjacent captured photons to jostle their positions to reestablish equilibrium around the entire surface of the photon shell to balance the forces, as shown in Figure 6-4. As the photon shell expands, the distance from the proton surface to the photon shell increases by a specific amount determined by the mass and charge of the captured photon.

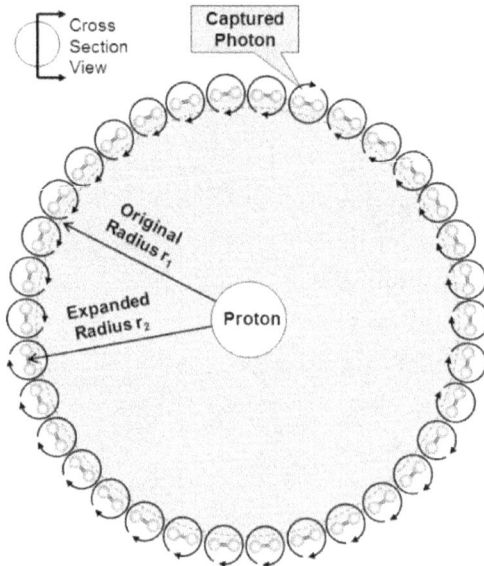

Figure 6-4. Photon shell diameter expands
to accommodate captured photon.

Now the photon shell is once again stable. If this event was the only photon absorbed with no change to the environment around the atom, then the photon shell would want to emit another photon to return to its previous or lower energy state (balance point), as depicted in Figure 6-5.

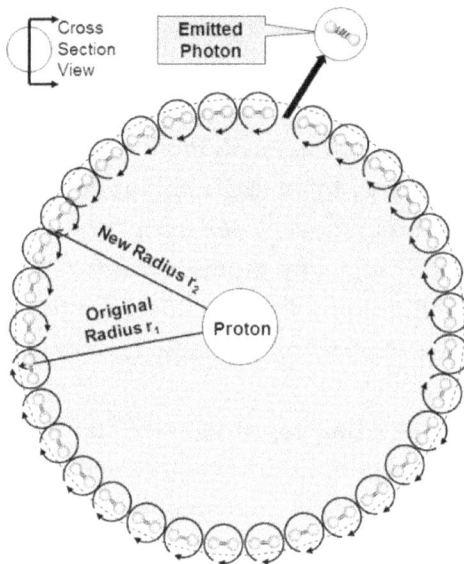

Figure 6-5. Photon Is Emitted from Photon Shell

Photon Emission

The mass and charge of the emitted photon is determined by the physical state of the atom and the environment around the atom. For example, if we heat[47] a sealed vessel[48] of hydrogen, then all the atoms in our sample would begin to absorb photons. It is the additional mass and charge of these absorbed photons that determine the physical state of the absorbing atoms. In this example, had we begun with solid hydrogen, we would see it first transition from a solid to a liquid, and then from a liquid to a gas, as each atom is forced to add more photons to its photon shell.

Conversely, if we remove the heat source and allow each atom to emit photons, we would see each atom revert from a gas to a liquid, and then a liquid to a solid.

---

[47] When we say "heat," we are really referring to an excess supply of photons made available to the atoms.

[48] In this example, a sealed vessel is just a physical containment for the atoms themselves. Photons can readily pass through the containment vessel to be either absorbed or emitted by the atoms within.

When a photon is emitted, there is a differential of mass and energy between the specific atom involved and its neighboring atoms. Thermodynamic experiments, such as those conducted by Sir Isaac Newton, measured that atoms with more mass and energy will always give up mass and energy to any adjacent atom that has less mass and energy. Therefore, we can easily see from these same thermodynamic principles that two adjacent atoms would automatically exchange photons in an attempt to nullify the difference between them. It is this atom-to-atom mass/energy relationship that determines the physical state of each atom.

Now let's look at what happens when the photon shell reaches a point where it emits a photon to lower its mass-energy state. Responding to the mass-energy differential among its adjacent atoms, the photon shell begins a rebalancing, which will eventually find the most massive photon and eject it from the shell. Why the most massive? Remember, the photon shell always wants to contract to its minimum radius. The largest photons represent the greatest discontinuity on the surface of a photon shell. This discontinuity results in a force imbalance, where the repulsive force of the atomic nucleus is concentrated on that single photon, and the lower mass photons around it will be unable to keep that larger mass photon as part of the shell.

When the photon separates from the photon shell, the electrostatic force between the atom and the photon immediately causes the emitted photon to accelerate away from the atom. As the photon gets further away from the atom, the gravitational imbalance diminishes, enabling the two TON particle objects inside the photon to resume vibration.

Meanwhile, back in the photon shell, another rebalancing effort begins, and the process repeats until the atom no longer experiences a mass and energy differential among its neighboring atoms. TON Particle Theory is therefore in full compliance with the laws of thermodynamics (mass/energy dynamics).[49] All energy and mass are conserved

---

[49] The exchange of photons clarifies that the laws attributed to thermodynamics apply to all mass-energy exchanges throughout the electromagnetic spectrum. Therefore, the term "thermodynamics" should be replaced with the term "mass/energy dynamics" in all but heat related engines.

during this process, just as Newton's law says, but we can now better define entropy, which is considered a measure of the unavailability of usable energy in a thermodynamic system.

Entropy

In thermodynamic theory, you have an "ideal engine" whereby energy is applied to the system. The engine does "work," and then the system returns to its initial state. Theoretical laws of thermodynamics identify that the energy applied to the system should equal the amount of work obtained from the system, if there are no losses. Experimental engines consistently defy this ideal model. Unable to account for the missing energy, scientists introduced the term "entropy" to define the missing energy[50].

Since all these experiments were only measuring photons in the thermal (infrared) regions of the spectrum, they missed the energy contained in photons at other spectral ranges. If we look beyond this narrow range of infrared frequencies, we will discover the emission of other photons and will therefore be able to account for all the energy and mass used for the (heating or cooling) process instead of relying on an undefinable "entropy" to account for the energy-mass discrepancy.

Plasma

Not all that long ago, we were told there are three states of matter: gaseous, liquid, and solid. Science now recognizes plasma as a fourth state of matter. Some theories speculate that, when enough excess mass and energy is added to an atom, the electrons (photon shell) and nucleus separate into ionized particles.

In TON Particle Theory, we can show that the photon shell in this higher mass/energy state *expands* to a point where photons are more easily absorbed and emitted. So, the current definition of plasma is half right. We are just clarifying that these expanded photon shells support the rapid exchange of mass and energy that enables atoms to

---

[50] Entropy is also considered a measurement of growing disorder in the universe. TON Particle Theory shows that this growing disorder does not exist.

rapidly change state from solid to liquid to gas.

Consider a plasma torch, which emits an electrical arc of photons to cut solid mass objects. When the atoms of the material to be cut absorb the photons emitted by the torch, the photon shells of the targeted atoms rapidly expand, which changes their physical state from solid to liquid and/or gas within a very short time. If the photon exchange area is small enough in size, you could theoretically cut the target material at the width of a single atom.

In classic thermodynamics, the electric arc emitted by the torch is considered to be in a plasma state. However, TON Particle Theory shows that at no point in time is the atomic integrity of either the "plasma" atoms or the atoms of the target material violated.

In theory, using a laser to cut a material works on the same principle. However, the laser uses photons at a carefully selected frequency, which are readily absorbed by the specific material being cut. The plasma torch, on the other hand, accomplishes the task by overwhelming the target material atoms with a wide spectrum of photon frequencies, of which only a handful are absorbed to inflate the photon shells.

The neutron has been considered a major particle in atomic theory, but is it as important as we believe?

Chapter 7. The Neutron

Now, let us consider the neutron. A classic neutron is approximately the same mass as a hydrogen atom, is approximately the same physical size as a proton, but does not exhibit any apparent charge.

The "coincidence" of similar mass between a hydrogen atom and a classic neutron is just too overwhelming to dismiss. TON Particle Theory presents the basic structure of the neutron object as a "failed" hydrogen atom, composed of a proton with a truncated photon shell pushed much closer to its surface than in a normal hydrogen atom.

Neutrons as "Failed" Hydrogen Atoms

In TON Particle Theory, a neutron would begin as a spherical collection of fused TON particles (a proton), which has attracted a photon shell in the star's photosphere and is on its way to becoming the hydrogen atom shown back in Figure 6-1. However, the hydrogen atom is then pulled back down into or near the fusion zone. Because there is a lot of space inside each photon and between adjacent photons in the photon shell, we hypothesize that gravity forces the photon shell closer to the surface of the proton. As the diameter of the photon shell decreases, electrostatic forces cause the photon shell to emit some of its photons. The result is a hybrid subatomic particle with slightly more mass than a proton, as seen in Figure 7-1.

The presence of captured photons close to the surface of the proton would explain its difference in measured electrical and magnetic properties. The captured photons are still intact and rotating in close proximity to the proton surface. The photons would form a magnetic field around the proton, insulating the charge and making the resulting TON particle object appear to be neutral in charge.[51] These characteristics match those the scientists have attributed to the classic

---

[51] For more details, see Proof 3, which mathematically verifies the magnetic insulation properties of the spinning, captured photons in the photon shell.

neutron.

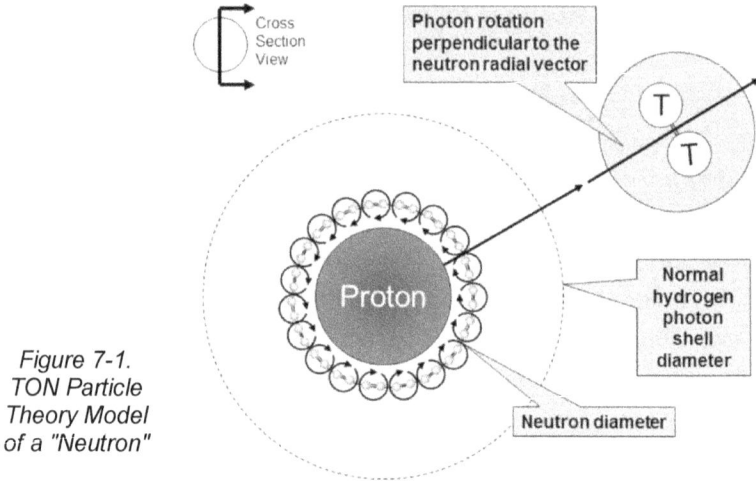

Figure 7-1.
TON Particle
Theory Model
of a "Neutron"

Using TON Particle Theory, we can account for the large number of variations in charge, permeability, permittivity, and conductivity scientists have observed and measured in neutrons. These variations are due to the different distances that could be formed between the photon shell and the proton inside. Variations in neutron mass can be explained by the fact that the photon shell of a hydrogen atom can include captured photons of many mass variants (Chapter 4).

Variations in the final distance between the photon shell and proton surface depends upon how much gravity the hydrogen atom experiences in the compression phase. Greater compression would result in a smaller diameter of the photon shell, which would cause more photons to be emitted due to electrostatic forces, resulting in lower mass.

TON Particle Theory's proposed method of neutron formation also accounts for observed neutron beta-particle emissions. Let's say a neutron (failed hydrogen atom) escapes the star's gravitational field into free space. Without the forces of the star, the neutron/atom returns to a state where it will try to balance the photon shell around the proton. However, after being compressed, the photon shell now may have insufficient mass to achieve stability. Under these conditions, the photon shell will expand to a point where the electromagnetic field generated by the spinning, captured photons becomes too weak to retain the integrity of the shell and it becomes unstable. At that point, the photons would all be emitted, leaving just the proton. There are many scientific papers describing the occurrence of these unstable neutrons emitting gamma rays, beta-particles, and other types of photons.[52] We can easily see how such variations can occur, as there are near infinite combinations of captured photons within the photon shell and the depths to which they can be pushed closer to the proton surface.

In summary, as we assess this new view of the neutron, we will begin to better understand the neutron and how we can exploit its features and participation in complex atomic structures.

---

[52] Classic physics describes this event as "neutron decay."

Why doesn't the Periodic Table[53] include deuterium and tritium? Read on and be enlightened.

Chapter 8.
Complex Atoms and the Periodic Table

The classic atomic model combines protons, neutrons, and electrons to form a stable atom. However, TON Particle Theory creates the same atom using only a series of fused proton objects and an expanded photon shell held together by simple electrostatic and gravitational force-balancing. (See Proofs 1, 2, and 3 for the mathematics involved.) We will now build upon this simple model to show how TON Particle Theory's stellar fusion model creates the rest of the elements in the Periodic Table and a few that are missing.

Let's start with deuterium. The current periodic table refers to deuterium as an isotope of hydrogen with a proton and a neutron (to account for the added mass in the nucleus) and one electron. In Figure 8-1 we show this model, substituting a photon shell for the classic single electron.

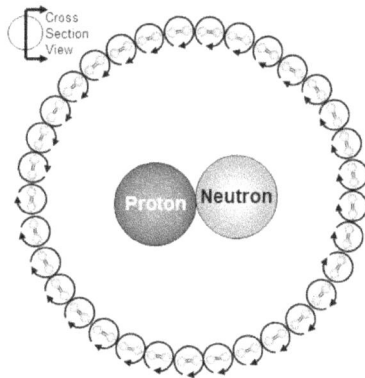

Figure 8-1.
The Classic
Deuterium Atom
(with photon shell)

The classic model results in an irregularly shaped nucleus, which we maintain would destabilize the photon shell, causing the sporadic

---

[53] The Periodic Table is a visual classification of chemical elements introduced by Dmitri Ivanovich Mendeleev (1823-1907) in 1869.

emission of photons. While astronomers have verified the existence of stable deuterium atoms in great abundance in comets, science has not found deuterium atoms that sporadically emit photons. Only a spherical nucleus can support a stable photon shell and, therefore, it is more likely that the deuterium nucleus is indeed a single, fused sphere with the mass equivalence of two protons. As explained in Chapter 3, such a single, spherical object with twice the mass of a proton would be a logical product of TON Particle Theory's stellar fusion. When combined with a slightly expanded photon shell, the result would be a stable deuterium atom (Figure 8-2).

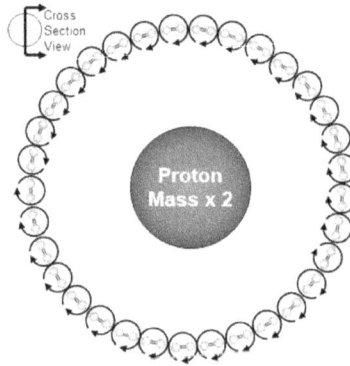

Figure 8-2. Deuterium Atom with a Spherical, Fused Nucleus and a Photon Shell.

The fused deuterium nucleus with its increased mass and charge requires an additional 0.26 electron mass equivalence in the photon shell to achieve balance between the electrostatic and gravitational forces. This is achieved in TON Particle Theory by capturing additional photons and/or photons of larger mass.

The increase in diameter of the deuterium nucleus would be very small, compared to a single proton. From geometry, we can calculate that the additional mass or volume of the second proton increases the diameter by about 26% over that of a single proton, since diameter changes with the cubed root of volume ($v^{1/3}$). Hence for a twofold (2x) increase in a nominal unit volume (1.0), we calculate $2^{1/3} = 1.26$ which is a 0.26 or 26% increase in diameter. This size difference is negligible given the distances involved.

This minor increase in atomic diameter agrees with the findings of Nobel Laureate Niels Bohr, who calculated the diameter of most stable atoms consistently stayed at about the same size at their lowest energy states. The now-called Bohr's radius establishes the approximate size of the photon shell around a stable nucleus. TON Particle Theory's atomic model confirms that Bohr's radius is the approximate distance maintained between a photon shell and the nucleus' surface. All of this is achieved with just a single layer of captured photons, eliminating the need for the multiple levels of electrons needed to support heavier atomic elements in the current Periodic Table.

Tritium

The next logical atomic nucleus created by the TON Particle Stellar Fusion model would be tritium, which is also missing from the classic Periodic Table. In classic physics, a tritium atom would be composed of one proton, two neutrons, and a single electron. (In Figure 8-3 we have again substituted a photon shell for the single electron.)

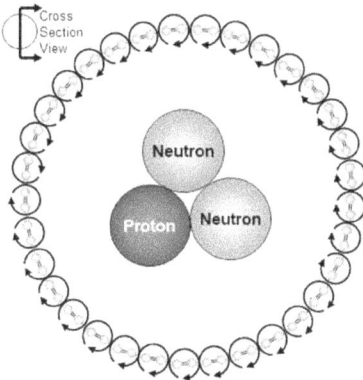

Figure 8-3. Classic Model of Tritium (with Photon Shell)

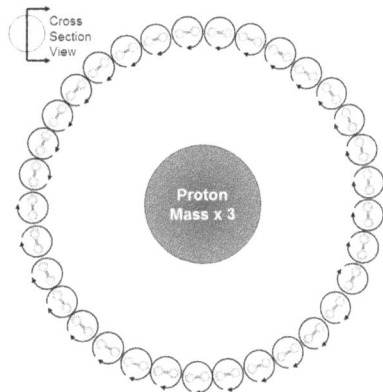

Figure 8-4. Tritium Atom with Fused, Spherical Nucleus and Photon Shell.

In TON Particle Theory, a tritium nucleus would be formed in a boundary layer fusion event[54] of a single proton with a single deuterium nucleus (Figure 8-4). TON Particle Theory has already established that the irregular nucleus of the classic view would be unstable and

---

[54] See Chapter 3.

incapable of supporting a stable photon shell. Since stable tritium atoms are commonly found in comets and on earth, it is more likely that these stable atoms contain a single, smooth, spherical nucleus with a mass equivalent to three protons. In addition to the additional mass in the nucleus, a stable tritium atom also requires additional and/or more massive photons in the photon shell to properly balance the gravitational and electrostatic forces and to maintain Bohr's radius. A quick review of geometry identifies that we need additional photons with electron equivalent mass of 0.44 over the hydrogen atom to achieve balance.

It is quite possible that deuterium and tritium were excluded from the classic Periodic Table because there were no theories at the time to account for their needing an additional 0.26 and 0.44 electron mass equivalents to balance the math. TON Particle Theory resolves this problem, enabling them to take their rightful place in the pantheon of atomic elements.

Helium

The atomic geometry gets very interesting when we consider the structure of a helium atom, as shown in Figure 8-5 and 8-6.

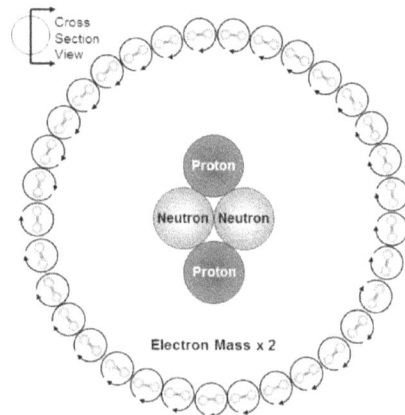

Figure 8-5. Classic Helium Atom with Heavier Photon Shell

The classic model of the helium atom has two protons and two neutrons in the nucleus, with two orbiting electrons. We have already described how TON Particle Theory replaces the idea of orbiting electrons with a photon shell (Figure 8-5). To balance the helium atom's nucleus, the photon shell would have to contain 1.52 electron equivalents. This is accomplished by adding additional and/or more massive photons to the shell. The next problem with the classic model is that it would create an unstable photon shell due to the irregular geometry of the nucleus (Figure 8-6).

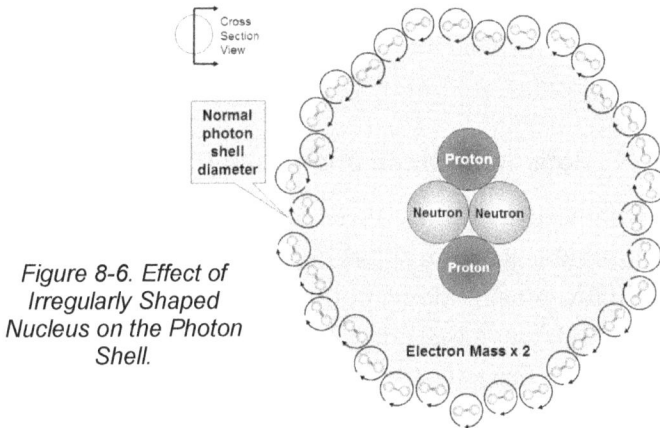

Figure 8-6. Effect of Irregularly Shaped Nucleus on the Photon Shell.

The final problem with the classic model is that science has not yet fully explained how to combine two protons and two neutrons to form the helium nucleus. Nor has it found a way for two electrons to share the same "orbit."

As discussed in Chapter 3, TON Particle Theory's stellar fusion model would normally create a single, spherical, fused object with the equivalent of four proton masses from two deuterium nuclei-sized objects. To form a stable helium atom (Figure 8-7), this object would need to capture a photon shell with the mass equivalence of 1.52 classic electrons, which is consistent with the geometry used for stable deuterium and tritium atoms.

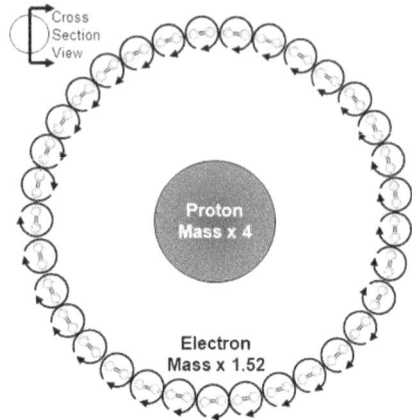

Figure 8-7. Stable Helium Atom with Spherical, Fused Nucleus and Photon Shell

Helium Isotopes and Alpha Particles

Classic physics identifies three helium isotopes, which are all considered unstable and have been documented as He3, He6, and He8. These isotopes most likely result from incompletely fused nuclei. As we have noted previously, the irregular shape would prevent an incompletely fused nucleus from forming a stable photon shell, and thereby a stable atom. It is also important to note that there is also evidence of stable atoms with the same mass as these isotopes, indicating that the stellar fusion process appears to work according to TON Particle Theory.

Figure 8-8. The Classic Alpha Particle (Helium-4 or He4 Nucleus)

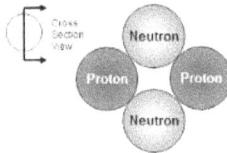

Figure 8-9. Alpha Particle in TON Particle Theory

Classic physics claims that the helium nucleus, stripped of its electrons, results in what has been called an alpha particle (Figure 8-8). TON Particle Theory explains that, if we had a partially fused configuration of four protons, they would inherently be unstable and therefore would not have a photon shell, which could explain the appearance of these alpha particles. The equivalent TON particle object would be a fully-fused, spherical object (Figure 8-9) capable of capturing and retaining a stable photon shell.

Nature abounds with examples of perfect symmetry creating stable structures. Atoms are no different—having a completely fused, spherical nucleus keeps the forces in balance and the atom structure stable. Irregular, partially fused nuclei result in unstable atoms, which are prone to losing their photon shells. So, we now have a clear scientific distinction between stable, non-radioactive elements and radioactive elements (see Chapter 11). It all comes down to having a smooth, spherical, atomic nucleus.

We could continue with these examples into the rest of the Periodic Table and the various isotopes, but we have made our point.[55] A star will fuse larger and larger mass objects into single, smooth spheres on a geometrical progression as long as it has the mass and gravity to do so. Stellar fusion continues until the star can no longer fuse the last sized objects it creates. At that point, the star begins to accumulate the largest objects it can fuse at its center. (See Chapter 12.)

---

[55] Look for our future book where we completely redefine the periodic table.

Can an entire electron particle flow from atom to atom?
We think not.

Chapter 9. Electricity or Charge-Carrying Photons

TON Particle Theory better explains the flow of charge in what we call electricity. The classic view has electrons ripped from one atom to another, yet there has been no mathematical evidence that such an event can occur. The forces and energy required to pull even a notional electron particle out of an atom and to insert that electron particle into an adjacent atom without destroying that atom has never been established. It has been assumed that the electrons flow due to a difference in what has been called the Electro Motive Force (EMF)—commonly measured as voltage—across the conductive material. The classic experiment applies a voltage across a conductor with the current measured in amperes. Again, it has been assumed that the current is composed of electrons moving, as that is all the classic model of the atom could provide as a mechanism for carrying charge, but there is no scientific experiment that has proven this process to be true.

Consider hydrogen gas in a glass vessel with two electrodes attached. Scientists have measured current flow through the hydrogen gas, while maintaining the integrity of the gas. If hydrogen was truly composed of a single electron and a proton, such an experiment should destroy the integrity of the hydrogen atom when the electron is removed. Yet, that does not occur. Clearly, something other than electrons are moving between the atoms.

Now, consider the case where we replace the single "orbiting electron" of the hydrogen atom with a photon shell composed of approximately 3.101×10^{32} unit photons. If we apply a voltage across the hydrogen gas now, individual photons can move from atom to atom, creating the same current flow without disrupting the integrity of the hydrogen atoms. The mathematics involved (Proofs 1, 3, and 4)

demonstrate that single photons are much easier to move than a particle the mass of an electron. The flow of photons, rather than electrons, also enables a much finer degree of low current transfer
through atomic materials, especially semiconductors.

At the atomic scale, when we establish a voltage difference or EMF
across a material, the flow of photons depends both upon the atomic
structure and, to a large extent, the crystalline/molecular structure
of the element being used to carry the effective electric charge. In the
case of a conductor, seen in Figure 9-1, we have photons moving from
one atom's photon shell to the next.

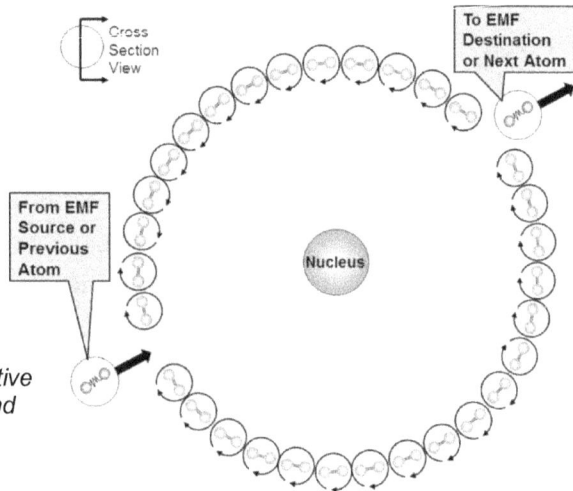

Figure 9-1. Conductive
Atom Captures and
Emits Photons

The EMF across each conducting atom facilitates photons moving
from atom to atom, flowing from the atom with the higher potential to
the atom with the lower potential, which satisfies current, or charge-
flow, conditions.

Impediments to Current Flow

If the conductor atoms are pure and homogeneous, then the photons can move current without loss across the atoms' photon shells.
However, if there are impurities (other elements) in the conductive
material, then there will be instances where the impurity absorbs the
charge-carrying photon without emitting a corresponding charge-carrying photon to the next conducting atom.

Usually, the absorbed charge-carrying photon causes the impurity to emit a different type of photon, typically a photon vibrating in the infrared (IR) spectrum, as previously seen in Figure 9-1. We know that this mass/energy transfer event happens, since nearly all our normal conductors have some resistance to current flow and emit heat (infrared photons) when carrying current.

Scientists have well documented that current flow often results in infrared emissions from impure conductors. Indeed, the impurities of some materials have been used to purposely convert normal current flow into specific emissions of photons in specific wavelengths (e.g., in the X-ray machine). The more impurities there are in the conductor, the more heat is generated and the less current flows through the conductor (resistance). For example, the Ni-chrome (nickel-chromium) wires in your toaster are specifically manufactured to exploit this fact.

Electric Insulation

In an electrically non-conducting material (insulator), the atoms involved are less likely to support the flow of photons. This non-conducting effect arises because some atomic elements have a different level of balance between the atomic nucleus and the photon shell at "room temperature." The difference in this force balance is directly related to the number of photons needed to change an individual atom from a solid to a liquid state. At room temperature, some atoms and molecular compounds will absorb large numbers of photons before the photon shell reaches an appropriate size to allow it to begin to exchange photons with neighboring atoms. However, if the EMF is large enough, even the best insulator will conduct charge.

A similar effect occurs when the adjacent atoms have a complex geometrical arrangement (a compound or a crystalline structure). For example, rubber has a complex molecular structure that tightly binds carbon and hydrogen atoms. The hydrocarbon compounds are so tightly bound together by the gravity and electrostatic forces between the atoms that they will not easily free up photons because of the more intricately constructed, mutual photon shell.[56] Even when large EMF is applied, the hydrocarbon structure will absorb the inflow of photons from those atoms with the highest EMF difference, but retain them instead of reemitting them as current. In this case, the atomic/chemical structure of the hydrocarbon molecule permits it to behave as an insulator. It is their photon absorption capability that permits those compounds to resist current flow and be successfully used as electrical insulation.

Super-conductors

In a super-conductor, the material is "cooled[57]" to its lowest energy state, which means that the resulting atom and/or compound has the minimum number of photons in the photon shell. In most atoms, this state is achieved when only the least massive photons (unit photon) remain in the photon shell. In this state, the distance between the photon shell and the nucleus is at its minimum, and the distance between neighboring atoms is also at minimum. This arrangement slightly increases both the gravitational and electrostatic forces needed to capture and/or emit new photons. In addition, since we are dealing with the photons with lowest mass, the amount of force needed to capture and/or emit them is also at its minimum. Therefore, any photon that approaches the photon shell of a super-conductive atom is quickly absorbed and another photon is emitted because, under these conditions, the photon shell has fewer photons and a smaller surface to rebalance. That is why super-conductors exhibit the highest electrical conductivity with a minimum of energy loss.

The Effect of Conductor Diameters

TON Particle Theory also explains why wire size has a major effect on current flow. Large diameter wires require less voltage for a given amount of current (photon transfer), because the larger number of

---

[56] Organic chemistry identifies that hydrocarbons form very compact structures. It is often pictured with the atoms sharing electron particles, which in TON Particle Theory would indicate that these atoms would share a common, single photon shell.

[57] In TON Particle Theory, the terms heating and cooling are artifacts of a thermodynamic view of the universe. The correct term is the absence or abundance of photons in the Photon Shell.

conducting atoms facilitates the movement of photons from one atom to the next. Small diameter wires reduce the number of conducting atoms available, thus limiting the number of photons that can flow from atom to atom, as shown in Figure 9-2.

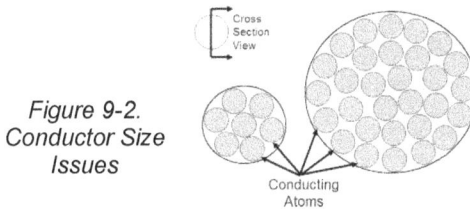

Figure 9-2.
Conductor Size
Issues

Crystalline Structure of Conductors

Another issue that affects current flow is the crystalline structure of most conductors, as seen in Figure 9-3.[58] A crystal has a very efficient packing factor, which places atoms very closely together and facilitates the photon exchange. We know from semiconductor properties that the crystalline pattern can assist or impede the movement of photons through the material, depending upon the atoms involved.

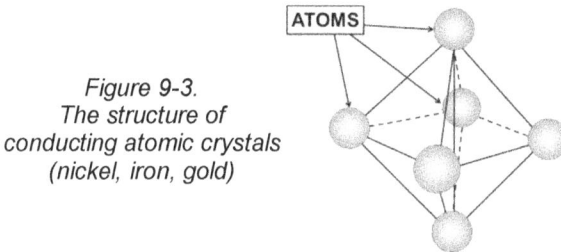

Figure 9-3.
The structure of
conducting atomic crystals
(nickel, iron, gold)

In rare instances, the formation of a crystalline structure impedes the photon exchange. As a liquid, water is an excellent conductor because its molecules can get very close together, so they can easily exchange photons between molecules. In its solid form (ice), the crystalline structure forces the water molecules further apart, making the ice less conductive than liquid water.

---

[58] This crystal issue may also resolve why some atoms are "ferro-magnetic." You will understand this supposition when you get to the chapter on magnets, Chapter 10.

Thermodynamics

TON Particle Theory's description of charge flow is also consistent with the flow of "heat" in thermodynamics.[59] The only difference between current flow and "heat" flow is the types (mass/charge and vibratory frequency) of photons moving from atom to atom. When we "heat" or "cool" a material, we are just reducing or increasing the number of photons in each atom's photon shell. With this new insight, we should be able to greatly improve the efficiency of heating and cooling systems. By better understanding how photons interact with various atoms, we will eventually be able to match the atomic materials to the correct wavelengths of the photons to maximize heat flow and thus greatly reduce the energy needed to accomplish the same tasks.

The Calorie and the Photon.

When looking at mass/energy transfer between atoms and compounds, the world scientific community uses the calorie as its main unit of measure. A calorie is defined as the energy needed to raise the temperature of one gram of water (H_2O) one degree Celsius at standard atmospheric pressure. The calorie is defined as 4.184 Joules of energy.

There are 3.346×10^{22} water molecules in one gram of water, which means that each molecule must absorb 1.2508×10^{-22} Joules of energy (EWM) to satisfy the equation. TON Particle Theory defines the unit photon[60] as having two single entangled TON particles with an estimated combined mass of 6.68×10^{-92} kilograms. The kinetic Energy of the Unit Photon (EUP) is:

$$\tfrac{1}{2}\ m(c)^2 \text{ or } 6.0811 \times 10^{-78} \text{ Joules.}$$

Where: m = mass and
c = speed of light

To calculate the Number of Unit Photons (NUP) needed, we divide

---

[59] See footnote 55.

[60] A Unit Photon represents the first photon created by a star. It involves just two TON particles and therefore represents the lowest mass photon possible. All other photons will have more massive TON particle objects within them and will thus represent larger increments of available mass/energy in a photon exchange.

the amount of Energy needed to raise a single Water Molecule by one calorie (EWM) divided by the energy represented by each Unit Photon (EUP).

$$NUP = EWM / EUP$$

So, the Number of Unit Photons needed to raise the temperature of a water molecule one degree Celsius is approximately:

$$2.0565x10^{55} = 1.2508x10^{-22}/6.0811x10^{-78}$$

We can use this estimate to begin assessing the number of photons exchanged in all thermodynamic equations that use the calorie as a unit of measure. We must remember that, until we can control the exact type of photons in each exchange, we should do full spectral analysis of each experiment to try to estimate the number of photons in each spectrum involved, as the real issue is the total mass and charge accumulated by each photon shell.

When Nobel Laureate Max Planck conducted his black body experiments in the early 1900s, his experimental apparatus was not capable of measuring the true spectral output of his heat source. Even he admitted that his now famous Planck Constant was merely a mathematical fudge factor used to fit his measured data. Now that we understand that the energy emitted from such a black body source covers a much larger range of spectra, his experiments and conclusions need to be reevaluated, as his constant is most likely incorrect.

If Maxwell's electromagnetic theory is correct, permanent magnets must have a source of charge flow.

Chapter 10. Magnets

The topic of magnets is one of our favorites, mostly due to the predictions of the monopole and magnetic domain theory. A monopole is a hypothetical particle that has only one magnetic pole, and has never been physically realized. The theory of magnetic domains has been extensively used to explain permanent magnets and has been reported to exist based upon crystals observed by the electron microscope. However, as we shall see, there is something else happening at the atomic level.

Let's get down to the real issue. Maxwell's equations and experiments show the relationship between current flow and magnetic field creation very clearly. To have magnetism, we must have charge flow. Yet, such things as permanent magnets exist, without any apparent charge flow.

How can different types of materials retain a magnetic field without apparent charge flow?

Here's the answer according to TON Particle Theory. Look at the crystalline structure of our best-known natural magnetic rock, the iron-nickel meteorite, illustrated in Figure 10-1.

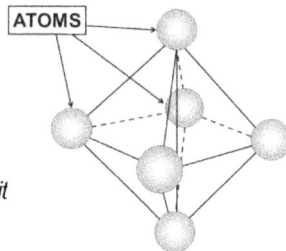

ATOMS

Figure 10-1.
The Iron-nickel Meteorit
Crystal Structure

It looks just like a conductor atom (Figure 9-3), doesn't it? This basic atomic structure facilitates the exchange of photons to support electric charge flow. It is this crystalline structure that suggests how the material could sustain a magnetic field.

As the meteorite entered Earth's atmosphere, it was "heated" to a molten state. As the meteorite "cooled" and re-solidified, it was exposed to the earth's magnetic field, which induced current flow around the short axis of the newly formed crystals, as shown in Figure 10-2.

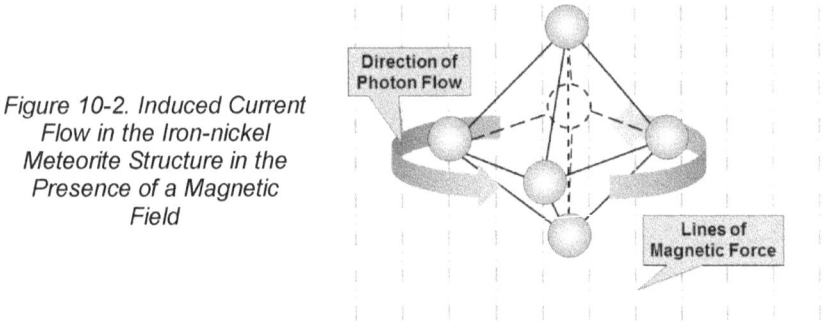

Figure 10-2. Induced Current Flow in the Iron-nickel Meteorite Structure in the Presence of a Magnetic Field

The charge-carrying photons move from atom to atom (photon shell to photon shell). In Figure 10-2 we show this as a counterclockwise flow, although it could be clockwise depending upon the orientation of the initial magnetic field. The moving photons induce a magnetic field perpendicular to the photon flow. Photons will continue to move around the short axis of the crystal, and the result is a permanent magnetic field running across the long crystal axis as shown in Figure 10-3. This photon flow will continue forever unless some outside force disrupts either the crystalline structure or the flow of the photons. This charge-carrying photon flow effect results in a permanently magnetic material down to the single crystal level, and thus creates the magnetic domains described in classic magnetic theory. An electron microscope would identify the larger crystals of magnetized atoms and, if they were tested individually, they would all behave exactly as the magnets themselves as long as the crystal matrix remained intact.

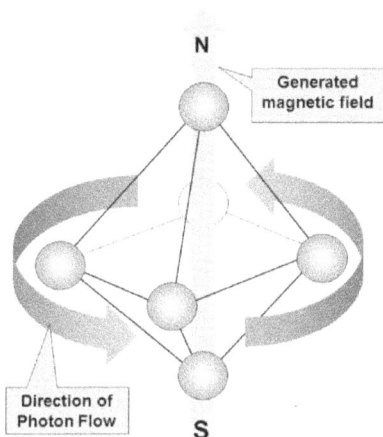

Figure 10-3.
Permanent, Induced
Current Flow in an
Iron-nickel Magnet

Once we realized the importance of this crystal formation, we did a check on all the "super magnet" materials and, sure enough, they all support the same or very similar crystalline structure. So, by adding aluminum and some of the rare-earth elements (alnico and neodymium magnets) to iron and nickel and melting them together, we find that they all fit tightly into the same structure. However, since they are different atoms, they cause the flow of photons between photon shells (current flow) to increase, and hence increase the magnetic field.

So, TON Particle Theory is the first theory to fully explain why some materials can retain permanent magnetic fields and others cannot. Those atoms considered "ferro-magnetic" are just those atoms that form the basic conductive crystal structure. The theory also fully proves that the theoretical "monopole" is not needed to account for permanent magnets, and that it violates Maxwell's equations and cannot exist.

Radioactive material is a gift from the universe. We should use it, not bury it as waste. It is far too valuable to us.

Chapter 11. Radioactive Atoms

We touched briefly on radioactivity earlier, but it deserves a bit more explanation. We earlier identified that an incomplete stellar fusion would create a non-spherical TON particle object. If this irregularly shaped object acquired a photon shell, the result would be an unstable atom. As discussed in Chapter 8, the classic helium atom, with its two protons and two neutrons in the atomic nucleus, would also create a non-spherical nucleus that could not sustain a stable photon shell. If we look at the resulting electrostatic and gravitational forces on the photon shell, we find that the forces are unevenly distributed among the surrounding photons. The uneven force distribution disrupts the spherical symmetry of the photon shell, as seen in Figure 11-1 (a repeat of Figure 8-6).

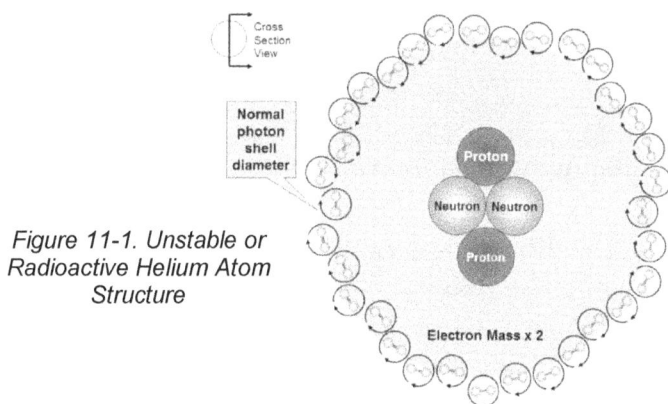

Figure 11-1. Unstable or Radioactive Helium Atom Structure

Yes, we have exaggerated the force effects in this figure, but we want to clearly illustrate the problem. If we have an irregular atomic nucleus, there is an inherent discontinuity in the forces affecting the photon shell. The force disruption results in sporadic events where a photon is emitted from the photon shell. Only a balanced, symmetrical photon shell will sustain a stable atom. A spherical nucleus comprised of proton-sized objects fused into one spherical mass supports that symmetry very easily. A non-spherical nucleus does not.

As an atom experiences normal photon exchanges, the photon shell will try to constantly adjust the photons around the nucleus, which will set up waves of photon position changes around its surface. In a non-spherical nucleus, it is only a matter of time before the surface waves synchronize into a "rogue" wave condition that tosses a photon out of the photon shell, as shown in Figure 11-2. Classic physics defines this event as a beta particle emission.

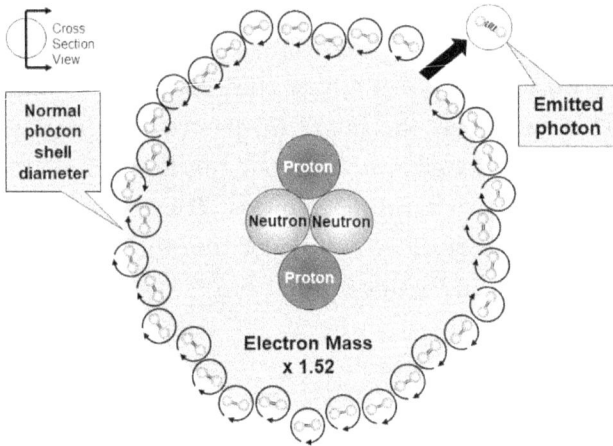

Figure 11-2. As a photon leaves the atom's photon shell, its two internal TON particle objects transition from rotation back to vibration.

Every radioactive element exhibits these photon emissions, and it is the interference of these photons with electromagnetic waves that led scientists to refer to these elements as "radio" active. Early scientific investigations concluded that the beta particles were equivalent to electrons. With TON Particle Theory, we can now see that these are photon emission events and, since photons come in many sizes, we can now better understand the different effects observed from these emissions.

When we look at a more massive radioactive material, such as uranium, we find a crystalline structure very similar to those we have identified in conductors and permanent magnets, as seen in Figure 11-3. The crystalline structure is key. As we have already seen when we looked at magnets (Chapter 10), it greatly facilitates photon flow—the transfer of photons from atom to atom along the short axis of the crystalline structure. In radioactive atoms, however, the irregularity of the nucleus and the balancing "waves" in the photon shell cause additional photons to be emitted in *random* directions.

Radioactive Half Life

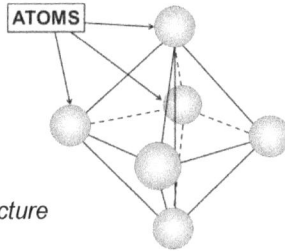

Figure 11-3.
Uranium Crystal Structure

To sustain photon emission, radioactive atoms need to continually replenish the photon shell with new photons to maintain the photon shell integrity. Most of the emitted photons will be absorbed by the surrounding atoms and provide them with the additional photons they need to sustain their radioactive emissions. Other emitted photons either escape to free space or are absorbed by non-radioactive atoms.

If the radioactive material is isolated from a source of new photons, then one-by-one each radioactive atom will cease to have enough photons to maintain its photon shell, and eventually the photon shell will disperse completely. Science has measured this reduction in the number of radioactive atoms in all radioactive material. They refer to the effect as the radioactive half-life.

Those atoms that lose their photon shells retain their intact atomic nuclei, but we do not currently have the technology to find them. The first problem in detecting these former nuclei is that they are incredibly small. Most of the volume of any atom is contained within the diameter of the photon shell, and the nucleus is a relatively miniscule part of that volume.

A perfect example of this differential can be represented by the relative size of the hydrogen atom and its proton nucleus. At Bohr's radius, the atom is approximately 10^{-10} meters in diameter. The proton (nucleus) is approximately 1.7×10^{-15} meters in diameter. So, the resulting nucleus is 60,000 times smaller in diameter than the photon shell. In the case of a uranium atom, the diameter of the uranium nucleus increases linearly, so the uranium atom, with the mass equivalent of 468 protons, would still be on the order of 10,000 times smaller than its corresponding photon shell.

The second problem in detecting former radioactive nuclei is that they no longer have any photons around them, so they give off no electromagnetic signature. This is why "dark matter" objects[61] cannot be detected using electromagnetic instruments.

Other Types of Radioactive Emissions

Earlier, we stated that radioactive "beta particle emissions" were initially considered to be emitted electrons. We have already established in Chapter 8 that electrons could not have been the source of these events. Based upon TON Particle Theory atomic model, these are photon events instead.

Radioactive gamma rays and X-ray emissions are also just different types of low-mass photons vibrating at very high frequencies. Their low mass enables both gamma rays and X-rays to more easily penetrate all but the densest materials before they encounter a photon shell to absorb them.[62] Scientists have well documented these high frequency photon emissions.

---

[61] "Dark matter" includes all TON particle objects that do not have a photon shell. Former radioactive nuclei are just one of the very large number of possible dark matter objects. See Chapter 2.

[62] Our description of these high-frequency photons runs counter to the energy predictions based upon Max Planck's radiation law. Now that we have established that photons have mass, it is clear that Max Planck's equation is probably wrong.

Symmetry is the foundation of the universe. The expansion and contraction of mass is just the heartbeat of the universe.

Chapter 12. The Big Crunch, or the End of the Universe as We Know It.

In Chapter 1, we identified the cause of the universe's expansion. Now we will show how TON Particle Theory explains how the universe will halt its outward expansion and then bring all its mass back to the point of origin to restart the expansion process in an endless universe life-cycle.

In Chapter 2, we showed how the universe initiates stellar fusion. The birth of the first stars begins the process by which the universe eventually eliminates the expansion force and enables gravity to dominate the universe.

Star Death

As each star fuses its TON particle objects, it builds progressively larger objects at its core as described in Chapter 2. The ultimate size of the TON particle objects a star is capable of fusing is dependent on its total mass.

Throughout its life cycle, a star also ejects huge amounts of variously sized and charged objects—photons, atoms, and dark matter—commonly referred to as the "solar wind." Eventually, the star loses so much mass/charge that its gravity is insufficient to continue primary stellar fusion[63] of the largest objects, and fusion ceases in the stellar core (Figure 12-1).

---

[63] We define primary fusion to be the ability of a star to create larger and larger TON particle masses, which converts kinetic energy into potential energy.

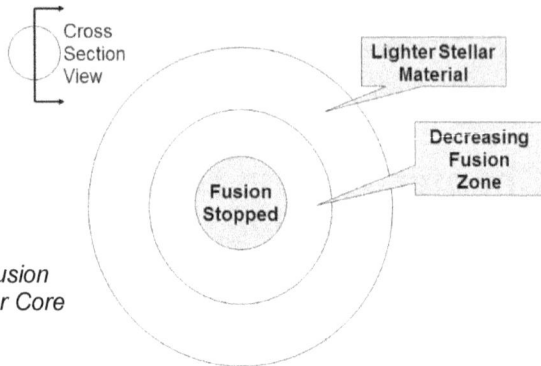

Figure 12-1. Fusion Ceases in Stellar Core

The stellar core continues to accumulate the most massive TON particle objects the star could fuse.[64] It is quite possible that the star retains sufficient gravity to force these large objects together so they are touching, but insufficient gravity to form fully fused spheres. This last point is important because, by touching, these objects become charge neutral between each other and therefore will remain together after the star collapses.

Even though the star no longer makes larger objects at its core, it does continue to fuse smaller sized TON particle objects in the fusion zones outside the core. We call this process secondary stellar fusion.

Stellar "Burnout"

Stellar "burnout" begins when gravity at the core can no longer push the massive TON particle objects together. As the gravity continues to lessen, each succeeding fusion layer, beginning with the one next to the core, either exhausts its source of fusible, equal-sized TON particle objects or can no longer continue the fusion process, as shown in Figure 12-2.

---

[64] These largest stellar objects could be anything from the size of a pebble to the size of a planet, depending on how much mass the star has to work with.

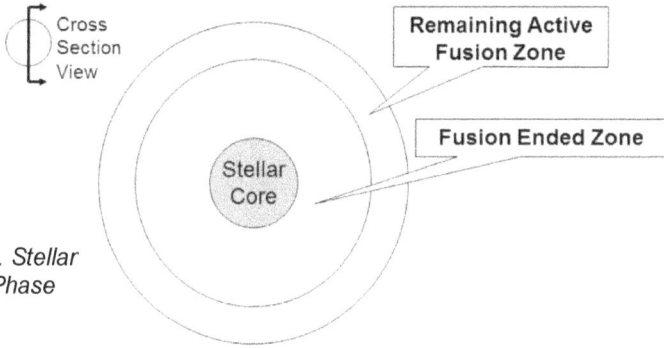

Figure 12-2. Stellar
Burnout Phase

As the fusion zones cease to function, the star continues to spew TON particle objects and photons outward, and eventually gravity is unable to support the balance between the electrostatic forces among the many types of fused TON particle objects. TON particle objects in each layer begin to repel each other, and some classes of stars begin to expand their physical size. (Astronomers call these stars "red giants" because, as the star expands outwards, the more massive photons (deep red and infrared) begin to accumulate in the photosphere to be expelled, making the star appear red in color.)

As mentioned in previous chapters, photons in the photosphere will spin instead of vibrate because stellar gravity compresses their spring constant (Chapter 2, Figure 2-6). Each photon then becomes a small electromagnetic toroid, and their combined magnetic fields create an overall photon magnetic field. Essentially, the star has developed one gigantic photon shell.[65] The star's photon interlocking magnetic field acts like a giant rubber band to maintain the integrity of the photosphere as it expands. That is one of the reasons why stellar expansion can continue over a very long time, even after most stellar fusion has ceased.

Ejection Sequence or Nova Event

As the star continues to lose mass to the solar wind, the force balance shifts from gravity-dominant to electrostatic-dominant. At that point, the star expels most of its outer layers, leaving just the most massive TON particle objects at its core. Since electrostatic

---

[65] Unlike the atomic photon shell, its depth will be greater than one photon.

forces are significantly stronger (10^4 times) than gravitational force, the outward mass acceleration increases at an exponential rate. That is why, when a star explodes, the event occurs very quickly. Astronomers describe this violent ejection as a nova event (Figure 12-3).

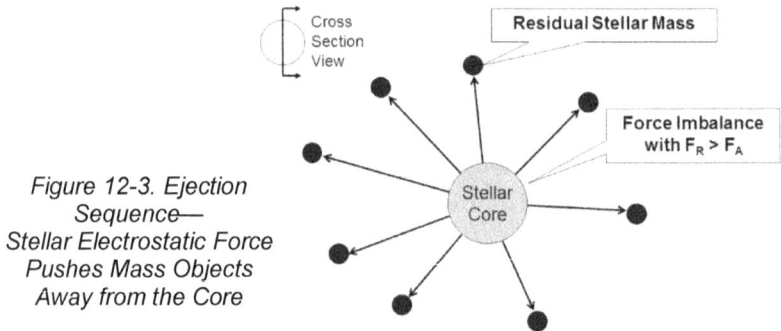

Figure 12-3. Ejection Sequence— Stellar Electrostatic Force Pushes Mass Objects Away from the Core

Some stellar models predict that the star will rapidly generate a series of heavy atomic elements at this point, but TON Particle Theory shows that the stellar fusion process formed those heavy TON particle objects before the ejection sequence (nova event) began. However, during the ejection sequence, the heavier, fused TON particle objects from deep inside the star do get mixed with the lighter TON particle objects (including photons) from the outer layers to form both stable and radioactive elements.

After the ejection sequence, all that remains of the star is its core, composed of the most massive fused TON particle objects the star could make, plus those TON particle objects that were pushed or pulled into the core during the ejection sequence. While some of these stellar core remnants are currently called "neutron" stars, this term is inaccurate. First, they can really be quite large—many times the size of a neutron and, second, there is no physical process by which neutrons can be grouped together, whereas the stellar fusion process will make large TON particle objects composed of the same material (TON particles) found inside a proton. The correct term for these solar remnants would more appropriately be small dense mass objects (SDMO).

The ejected star material remaining after the nova event will begin to re-collect around the stellar core remnant due to its local gravity, which sometimes leads to another stage of stellar fusion or the formation of what will eventually become a large dense mass object (LDMO) or, as it is commonly and mistakenly referred to in current astronomy and physics, a "black hole." The gravity of each SDMO will attract any smaller mass object and slowly increase its mass over time to transition to an LDMO. Eventually, they will get massive enough to fully fuse these additional TON particle objects into a spherical core or singularity.

LDMOs and Why You Cannot See Electromagnetic Emanations from Inside a "Black Hole"

When a photon enters the gravity field of a LDMO (i.e., crosses the event horizon), it accelerates towards the surface of the object. As the photon gets nearer to the surface, the LDMO's gravity compresses the photon's spring constant, the photon ceases to vibrate, and its constituent TON particle objects will begin to spin around each other.[66] This state change is the same phenomenon we have described whenever a photon interacts with a local gravitational field, whether that of a star or an atom.

Then, the electrostatic force of the LDMO forces the photon to spin such that the two TON particle objects in it are spinning perpendicular to the LDMO surface. Since gravity is dominant, the spinning photon continues to fall until it impacts the surface. At impact, the two TON particle objects that compose the photon disintegrate and evenly spread their individual TON particles across the surface of the LDMO like jelly over the surface of bread.[67]

Photons approaching the LDMO cannot be detected once they cross the event horizon because they cease to vibrate. Nor can we detect them after impact, because they no longer exist as photons, only

---

[66] When the photon begins to spin, it no longer generates an electromagnetic signature or vibrational frequency and is therefore undetectable by current technology.

[67] At the time the photon impacts the surface, the spring constant formed inside the star no longer exists, as it was tied to the center of mass of the two TON particle objects, which no longer exist as a unit.

as individual TON particles. In fact, we will not detect any electromagnetic radiation from the LDMO itself because its gravitational force is too great for it to emit detectible photons. To paraphrase the old roach trap advertisement, photons check in, but they don't check out.

Every TON particle object that impacts the surface of the LDMO contributes its mass/charge and has its kinetic energy converted back into potential energy as its individual TON particles fuse to the surface of the LDMO. The LDMO, then, acts like a giant vacuum cleaner picking up all stray TON particle objects that its gravity can reach. This "garbage collection" is the process that will eventually restore the universe to its point of origin.

The "Big Crunch"

Over time, each LDMO grows in mass and converts more and more TON particle kinetic energy into potential energy. As each layer forms on the LDMO, the conductivity and permittivity of each individual TON particle absorbs the charge of the objects inside, leaving only the outermost electrostatic charge able to affect any incoming objects. This means that, as the LDMO gets larger, its electrostatic repulsive force remains relatively constant over the surface area. This relative reduction in electrostatic force enables gravity to have an ever-increasing effect.

As LDMOs increase in mass and number, they will eventually consume galaxies and all stray matter between them and, eventually, they will begin to pull at each other. Observed as a system, the universe will begin to slow its expansion, stop momentarily, and then begin to contract. Slowly at first, then with increasing acceleration, the universe mass all comes home to its origin. No mass escapes and all past events and distances are irrelevant. The "crunch" is on. So, our TON Particle Theory predicts a complete life cycle for the universe, as shown in Figure 12-4.

In the end, we reach a new Time Zero, where all mass is bound together once more. Then everything begins again, as we described in Chapter 1. All mass and energy has been conserved throughout this process, as Sir Isaac Newton predicted.

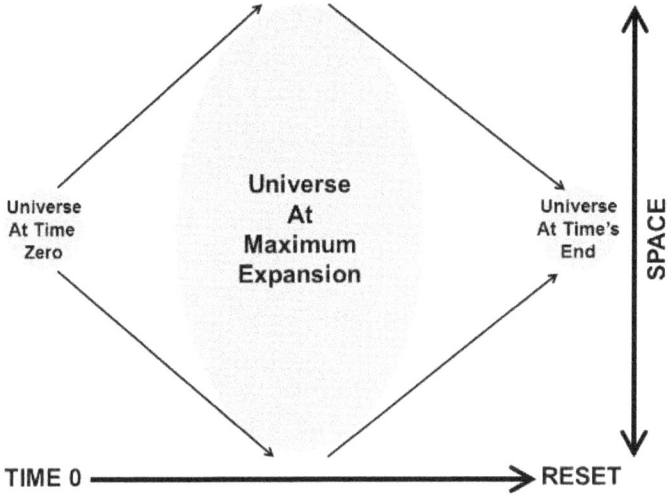

Figure 12-4. The Universe Life Cycle

Once you eliminate the impossible, whatever remains, no matter how improbable, must be the truth.
—Arthur Conan Doyle

Conclusion

TON Particle Theory brings order and understanding to our universe with a simple explanation of issues that current quantum mechanics and string theory have failed to resolve. Using one simple universal material and three well-known forces, in four well documented dimensions; we have shown how the universe could work. We introduce a new universal model that fully explains the expansion and contraction of the universe. We introduce a new stellar model that fully explains how stars convert the undifferentiated mass of the universal sphere into complex objects (atoms), and contribute to the slowing of the expansion. Our new atomic model cleanly explains all atomic attributes and eliminates a wealth of proposed subatomic particles and forces using very simple geometry and mathematics. This new atomic model fully supports most of the physics, chemistry and biology measurements and observations. We also present this list of current issues cleanly resolved by these new models.

Dark Matter — Fully explained with normal forces. No "Dark Energy" is needed
Radioactive Matter — Fully explained by non-spherical nuclei
Nuclear Fusion — Fully explained as an endo-energetic reaction. No energy breakeven point is possible.
Nuclear Fission — Fully explained as the only way to release atomic energy by converting potential to kinetic energy.
Gamma rays, cosmic rays, X-rays, ultra-violet, visible light, and infrared — Fully explained as to source, mass, and energy issues. They are not massless.
"Big Bang" or Great Expansion — Fully explained with all mass and energy conserved
"Big Crunch" — Fully explained with all mass and energy conserved
Star Formation — Fully explained by currently documented forces.

Star Burnout — Fully explained by currently documented forces.
Electricity — Fully explained by photon emission and absorption
Magnetism — Fully explained as photon mass/charge movement within a crystalline structure. There are no "monopoles."
Stable atomic structure — Fully explained with a fused, spherical nucleus and a single spherical photon shell
Thermodynamics — Fully explained as photon exchange with entropy resolved as a failure of spectral information capture
Speed of light — Fully explained as the time it takes two atoms to exchange a photon between their photon shells.
Photons — Fully explained as vibrating or spinning TON particle object pairs with mass and charge
Massless energy — Abolished as not possible
Mass and energy conservation — Confirmed in all reactions.

We believe TON Particle Theory is simple enough to work naturally, and nearly all our observations of the universe and atomic structures are easily explained with its basic models. TON Particle Theory also explains a lot of issues that classic and proposed theories cannot. It cleanly explains the universe from Time Zero to present day and shows a clear cyclical process, which predicts how the universe as we know it will continue to expand for a while before it reverses its expansion and ends back at the point of initial expansion.

Most importantly, the theory satisfies Einstein's criteria for a unified field theory. TON Particle Theory works at the extreme micro level of subatomic matter and at the macro (universe) level. If we can answer everything with one theory, then perhaps it is the ONE Theory.

Thank you for your time.

Glossary

Term, Abbreviation, or Acronym	Description
Beta particles	Classic physics defines these as "free electrons." TON Particle Theory defines them as photons emitted from a radioactive element.
Big Bang	Common name for the "beginning" of the universe, though it is just part of a repeating universe life cycle with no "Bang." Should more accurately be called the Great Expansion.
Black hole	Should be called a Large Dense Mass Object (LDMO), as it is not a hole.
Boundary Fusion	Fusion of TON particle objects with different sized masses
Cosmic ray	A photon vibrating between the frequencies of 10^{17} and 10^{18} Hz.
Dark matter	A TON particle object which is neither entangled with another TON particle object to form a photon, nor surrounded by a photon shell
Dark Star	A stellar body that either has not yet created detectable photons or lacks the mass to begin the fusion process
Energy	We are specifically referring to kinetic energy (mass in motion) and potential energy (mass at rest).
Entanglement	The process by which two TON particle objects are linked together by gravity and electrostatic forces.
Entropy	Measure of the unavailability of energy in a thermodynamic system, also, more generally, a measure of randomness or instability of a system. It is NOT a measurement of universal disorder.
ESA	European Space Agency
Fusion	The act of combining TON particle objects together into a single spherical object within a star.
Gamma ray	A photon vibrating between the frequencies of 10^{18} and 10^n Hz, where n > 30
Infrared light	Photons vibrating between the frequencies of 10^{12} and 10^{14} Hz.
Kinetic energy	Energy of a body (mass object) in motion

Term, Abbreviation, or Acronym	Description
LDMO	Large Dense Mass Object (aka "black hole" or singularity) created after a star completes its fusion cycle.
Magnetic Vessel	A notional space encapsulated by a magnetic field
NASA	National Aeronautics and Space Administration
Neutron	In TON Particle Theory, the neutron is a failed hydrogen atom and unstable.
Neutron Star	In classic physics, a notional stellar body with neutron-like particles as its sole mass component. Since the neutron we describe could not retain its structure in such a dense mass object, TON Particle Theory redefines this stellar body as a Small Dense Mass Object (SDMO).
Photon	Two TON particle objects of similar mass and charge that vibrate or rotate in a gravitational and electrostatic entanglement.
Photon Shell	A spherical collection of captured photons that replaces the notion of orbiting electrons in the atomic elements.
Photosphere	In classic theory, the visible, luminous area around a star, comprised largely of ionized gases and particles. TON Particle Theory redefines the photosphere as the collection of photons and TON particle objects in the outer layers of a star.
Plasma	In classical theory, matter consisting of highly energized, freely moving ions and electrons. TON Particle Theory redefines plasma as atoms with a super abundance of photons in the photon shell.
Potential Energy	Energy of a mass object at rest
Proton	A spherical collection of TON particles used as a basic atomic building block $\approx 10^{54}$ TON particles
Protostar	In TON Particle Theory, a collection of TON particle objects sufficient to initiate TON particle entanglement and fusion
SDMO	Small Dense Mass Object, a stellar core remnant after a nova event. This term replaces "neutron star" in all instances.
Singularity	See LDMO

Term, Abbreviation, or Acronym	Description
STAR	Simple TON particle Accretion Reactor. A stellar body that has created unit photons, beginning the stellar fusion process.
TON particle	The Only Needed particle, the basic universal material
TON particle object	Any single mass object made of TON particles that are touching each other.
Ultraviolet light	Photons vibrating at frequencies between 10^{15} and 10^{17}
Unit photon	A photon composed of two entangled TON particles
Visible light	Photons vibrating at a wavelength between 0.4 and 0.7 microns
X-ray	Photons vibrating at frequencies between 10^{17} and 10^{20} Hz.

Appendix: TON Particle Theory Proofs

In this book, we have journeyed into an exciting place, where we explore the existence, behavior, and relationships among subatomic particles we call "TON particles." Our aim was to introduce these TON particles, which are many, many orders of magnitude smaller than atoms, protons, neutrons, or electrons, and explore their role as THE basic particles underlying all matter and energy in the universe.

An important aspect of our TON Particle Theory, covered in the first several chapters, is how TON particles can cluster together, rotate, and vibrate with one another to form photons, dark matter, protons, neutrons, and atoms.

To verify TON Particle Theory, we recognize that we must provide a mathematical rationale that describes these behaviors and demonstrates that the theory is consistent with currently accepted measurements and equations provided by Einstein, Newton, Coulomb, and others. The results of our mathematical analyses were presented as the first three proofs in our original book[68], with some minor editing here. After further analysis, we now add two new proofs in this second edition/rewrite. Altogether, the five proofs address the following aspects.

Proof 1: A Mathematical Basis for TON Particle Proximity, or how TON particles can live together in harmony at very close distances, while satisfying the laws of Newton and Coulomb;

Proof 2: TONs in Oscillation, or how TON particles interact with one another in a dynamic oscillation or vibration, and still be consistent with accepted laws of motion (photon creation);

Proof 3: Photon Magnetic and Electrostatic Balance in the Photon Shell, or how charged TON particles in motion can create and maintain electrostatic and magnetic field balance within the atomic nucleus and photon shell;

Proof 4: More Concepts of Atomic Nucleus-to-Photon Shell Force Balance, or how balance between the photon shell and the

---

[68] *I Killed Schrödinger's Cat* by Donald A. Bertke and Herbert L. Hirsch. 2014

atomic nucleus may be attributed to electrostatic or electromagnetic force, expanding concepts developed in Proof 3; and

Proof 5: Atomic Stability in TON Particle Theory, or how the physical (not atomic) fusion of sufficient number of proton-sized TON masses into an atom's nucleus may make it appear stable, even though the nucleus is not a single fused sphere.

Some people may take exception to our choice of the word "proof," as it deviates from the pure mathematical definition of a proof, but we're not here to argue about semantics. We use the concept of mathematical proofs here to describe the mathematical models used to demonstrate how TON Particle Theory works in the real world.

Our proofs begin with previously defined and accepted laws of physics, which we use to *prove the plausibility* of TON Particle Theory. Since we are dealing with objects that our current technology cannot yet directly measure, physical proof can be inferred from currently available scientific data. TON Particle Theory provides a better explanation of the claimed results of many past scientific experiments.

TON Particle Theory provides the insight that should enable scientists to find and document the physical objects we describe. Researchers with the appropriate equipment may conduct several of the experiments, which we suggest. The resulting data will verify parts of our theory. Once science has accepted the existence of TON particle objects, this theory will enable them to better exploit this knowledge in all scientific disciplines.

The reader should note that the equations in these proofs are numbered sequentially within each proof, and not continually throughout the appendix, so we have an Eq.1, an Eq.2, and so forth in each proof.

The reader should also note that the currently accepted model of a number of electron particles orbiting an atomic nucleus, sometimes in many layers, is replaced in TON Particle Theory by a single layer (shell) of captured photons. We use the term "photon shell" to replace the orbiting electron model. Our proofs show that our photon shell model supports all the characteristics attributed to the older electron model. Our proofs also show that our photon shell model explains all mass and energy exchanges between atoms and is fully consistent with observations and measurements to date in all physical, chemical, and biological processes.

Proof 1: A Mathematical Basis for TON Particle Proximity

To estimate physical properties, such as mass, size (diameter), and electrical charge of a TON particle, we began with some fundamental physical laws and proven properties. Then we attempted to resolve these properties by (1) applying fundamentally sound mathematics, (2) remaining consistent with physical laws, and (3) not doing anything silly, like inventing alternate universes to explain things that did not happen to "fit" our ideas. The fundamental question we needed to confront was, "Is it plausible that like-charged particles, TON particles in our case, can accumulate in close proximity to each other despite the typically-dominant repulsive forces described by Coulomb's law?"

First, let's look at some fundamental laws of physics relating to gravity and electrostatic forces. We begin with gravity. Newton's law states that two masses m_1 and m_2 at a distance or range of r from one another exert a gravitational force of attraction, F_G, as:

$$F_G = \frac{Gm_1m_2}{r^2} \qquad [Eq.\,1]$$

Where: F_G = Force, in Newtons (N)
  m_x = Mass, in kilograms (kg)
  r = Range or distance between centers of mass, in meters (m)
  G = Newton's gravitational constant
    = 6.674 x 10^{-11} (N-m^2/kg2)

Similarly, Coulomb's law of electrostatic attraction or repulsion states[69] that two bodies with electrical charges q_1 and q_2 at a distance or range of r from one another exert an electrostatic force of attraction (if oppositely charged) or repulsion (if similarly charged), F_E, as:

$$F_E = \frac{Keq_1q_2}{r^2} \qquad [Eq.\,2]$$

Where: F_E = Force, in Newtons (N)

---

[69] *In TON Particle Theory we believe that electrostatic force is always repulsive, not attractive.*

q_x = Electrical charge, in Coulombs (C)
r = Range or distance between charge centers, in meters
K_e = Coulomb's electrical attraction constant
= 8.987x10^9 (N-m^2/C^2)

If we believe this phenomenology operates in some medium other than free space, K_e will be different. For now, we'll use the K_e value stated above, which assumes the permittivity of free space.

Now, let's say we have two bodies of similar and equal charge, q, and mass, m. For a certain charge/mass ratio, at any range, r, they will be in equilibrium – where the gravitational force of attraction equals the electrical force of repulsion, or:

$$F_G = F_E$$

In this situation, substituting the expressions for these forces from the two equations above, we have:

$$\frac{Gm_1m_2}{r^2} = \frac{Keq_1q_2}{r^2}$$

In our case, we are considering two TON particles of equal charge and mass, $m1 = m2 = m$ and $q1 = q2 = q$, so this becomes:

$$\frac{Gm^2}{r^2} = \frac{K_eq^2}{r^2}$$

Then, since range, r, cancels, we re-arrange the formula to get:

$$\frac{q^2}{m^2} = \frac{G}{K_e}, \qquad and \qquad \frac{q}{m} = \sqrt{\frac{G}{K_e}}$$

Now we have a charge/mass ratio expression for two bodies in equilibrium. These bodies could be planets, refrigerators, or anything. But for our purpose, let's say they are TON particles, adjust our notation accordingly, and calculate the numerical value using the values for G and K_e stated earlier:

$$\frac{q_T}{m_T} = \sqrt{\frac{G}{K_e}} = \sqrt{\frac{6.674x10^{-11}}{8.987x10^9}} = 8.617x10^{-11} \; C/Kg \qquad [Eq.\,3]$$

Where: q_T = Charge of a TON particle (in equilibrium with another TON particle) and
m_T = Mass of a TON particle

At this point we have two possibilities to consider. The first is that
when TON particles accumulate into a proton, or into any aggregate
mass for that matter, the total charge is distributed over the whole
volume of TON particles within the proton—a charged volume. The
second is that, when TON particles accumulate into a proton or larger
mass, the total charge is distributed only over the surface of the proton – a charged surface. We consider both possibilities below, in the
context of a proton.

Charged Volume. This possibility is illustrated in Figure P1-1,
and presumes some large number of TON particles are packed together to form the spherical proton. This illustration is conceptual and
not intended to be at scale, as there are many, many more TON particles in a proton than depicted here. All TON particles carry some
charge which must aggregate to form the total charge of a proton.

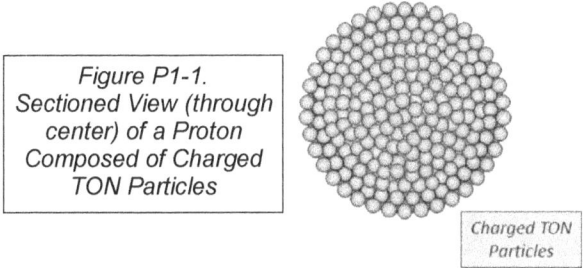

Figure P1-1.
Sectioned View (through
center) of a Proton
Composed of Charged
TON Particles

Charged TON
Particles

First let's see how much charge some number, n, of TON particles
in equilibrium may produce when they accumulate into a proton. We'll
call this a proton-equivalent charge, q_{PE}. Since we know the mass of a
proton, m_P, we simply multiply Eq.3 by that proton mass, which gives
us:

$$q_{PE} = \frac{m_P q_T}{m_T} = (1.672x10^{-27})8.617x10^{-11}$$

$$= 1.441x10^{-37} \; coulombs \qquad [Eq.\,4]$$

But we know that the actual charge of a proton is $1.602x10^{-19}$
Coulombs. Hence, we have charge multiplier (*CM*) effect where the accumulated charge of TON particles into a proton-equivalent mass is

actually less than the known charge of a proton, or:

$$\frac{q_P}{q_{PE}} = CM = \frac{1.602x10^{-19}}{1.440x10^{-37}} = 1.111x10^{18} \qquad [Eq.5]$$

This is an interesting effect, which says that the accumulation of some number of TON particles into a proton produces an increase in charge $1.111x10^{18}$ greater than the increase in mass. How do we explain this? First, since TON particles are, in our view, the most primitive particle, we'll say that the act of accumulating them does not change their mass or volume – they remain the same in terms of these properties through the act of accumulation. So, an accumulation of some number of TON particles, n, into a volume or mass equal to that of a proton - a *proton-equivalent volume*, v_{PE}, or *proton-equivalent mass*, m_{PE}, would be simply expressed as n multiplied by the mass or volume of a TON particle:

$$nm_T = m_{PE} \quad and \quad nv_T = v_{PE}$$

This effect would not explain the charge multiplication, so let's consider the possibilities. If charge accumulates disproportionately to mass or volume, it must accumulate proportionally to something else – some other physical property. After mass or volume, the next property that comes to mind is surface area. This would be a rational hypothesis, as there are other instances in electro-physics where charge is proportional to surface area, such as a hollow conductor. So, let's hypothesize that an accumulation of TON particle surface areas, where the physical accumulation is proportional to volume or mass, produces a surface area accumulation which, when divided by the surface area of a proton, is equal to the charge multiplier, *CM*, mentioned earlier in Eq.5. To test this hypothesis let's first think about how TON particles may physically accumulate.

To begin considering the physical accumulation, we'll first look at how spheres accumulate into a cube of some volume. First, consider a large cube of equal x, y, and z dimensions of X, and smaller cubes of some dimension, x, smaller than X. Then the number, N, of small cubes which can fit into the large cube is:

$$N = \frac{X^3}{x^3} \qquad [Eq.6]$$

Now let's move from cubes to spheres, in a semi-empirical way. The volume of a sphere of some general diameter, d, divided by the volume of a cube whose side dimension is also d, is:

$$\frac{Volume\ of\ a\ sphere}{Volume\ of\ a\ cube} = \frac{\pi d^3/6}{d^3} = \frac{\pi}{6} = 0.5236 \qquad [Eq.7]$$

So, we can multiply the number of cubes in a large cube by the coefficient of Eq.7 and obtain the number of small cubes in a large sphere. Now, let's say the small cubes represent the lattice structure of spherical TON particles packed into a large sphere, which represents the spherical volume of a proton. Then, the number, N, of Eq.6 becomes the number of TON particles in a proton, n, and X and x become the respective diameters of a proton and a TON particle:

$$n = 0.5236 \frac{d_P^3}{d_T^3} \qquad [Eq.8]$$

Here, we need to say a few words about spheres, lattice structures, smoothness, and packing factors. Eq.8 presumes a cubic lattice, which allows us to move directly from small cubes to small spheres, assuming each sphere touches six other spheres. A TON particle is a spherical entity and so is a proton, with some surface irregularity due to the clustering effect of all the spherical TON particles. A proton is a spherical accumulation of much smaller spherical TON particles. Since we presume a large number of TON particles per proton, we can think of a proton as being essentially a smooth sphere for our mathematical purposes here. The number of small spheres (TON particle volume) that can be "packed" into a large sphere (proton volume) is a function of how they arrange themselves into a lattice structure. Here, the density of the small spheres is defined by a packing factor, F_P. A cubic packing factor is, once again, the ratio of a spherical volume of diameter, d, to that of a cube whose side dimensions are also d, and yields a packing factor equal to the result of Eq.7, or 0.5236.

Gauss[70] and other scientists have calculated more efficient or denser packing factors, up to around 0.74 or so. For our purpose, we'll use the cubic factor, assuming the equilibrium balance amongst TON particles will hold them in a cubic lattice state. For now, we'll also assume no space between TON particles, and leave more quantified resolution of this distance to later analysis. If we were to choose a different packing factor, we would simply need to multiply the value of n in Eq.8 by that packing factor and divide the result by 0.5236 to correct that calculation for the chosen packing factor. We could also perform a more deterministic, trigonometry-based calculation of n for a chosen packing factor, but the empirically-derived value will suffice for now.

So, now we have an expression for n in terms of TON particle and proton diameters. We can also express the ratio of the proton-equivalent accumulation of TON particle surface areas and the proton surface areas. This, finally, allows us to calculate TON particle diameter. First, let's express the ratio of the accumulated sum of the number, n, of TON particle surface areas, a_T to the surface area of a proton, a_p:

$$\frac{\sum a_T}{a_p} = \frac{n\pi d_T^2}{\pi d_p^2} = \frac{n d_T^2}{d_p^2}$$

And since:

$$n = 0.5236\frac{d_p^3}{d_T^3} \qquad [from\ Eq.\ 8]$$

Then:

$$\frac{\sum a_T}{a_p} = \frac{0.5236(d_p^3/d_T^3)d_T^2}{d_p^2} = 0.5236 d_p/d_T \qquad [Eq.\ 9]$$

Our hypothesis is that the charge multiplier is the surface area ratio from Eq.9. This is a simplification because the charge multiplier is actually a ratio of surface area/volume quantities for accumulated TON particles and a proton. But volume is proportional to mass, and a proton-equivalent mass of TON particles is equal to a proton's mass. Therefore, a proton-equivalent volume of TON particles is equal to a

---

[70] https://en.wikipedia.org/wiki/Sphere_packing

proton's volume. Hence the volumes in the surface area/volume quantities would cancel and Eq.9 holds true. Continuing, and inserting the charge multiplier, CM, we have:

$$CM = 1.111x10^{18} = 0.5236d_p/d_T$$

Re-arranging, we can calculate the diameter of a TON as:

$$d_T = \frac{0.5236(1.7536x10^{-15})}{1.111x10^{18}} = 8.261x10^{-34} \ meters \qquad [Eq.10]$$

Then, it follows that the number of TON particles in a proton is:

$$n = 0.5236\frac{d_p^3}{d_T^3} = 5.0079x10^{54} \qquad [Eq.11]$$

And the mass of a TON particle is the mass of a proton divided by n, or:

$$m_T = \frac{1.6726x10^{-27}}{5.0079x10^{54}} = 3.3399x10^{-82} kg \qquad [Eq.12]$$

And the charge of a TON particle is the charge of a proton divided by the charge multiplier from Eq.5 to get the proton-equivalent TON particle charge, and then divided by n to get the charge of a TON particle, or:

$$q_T = \frac{q_P}{(CM)n} = \frac{1.6026x10^{-19}}{(1.111x10^{18})(5.0079x10^{54})}$$

$$= 2.8782x10^{-92} Coulombs \qquad [Eq.13]$$

We can do a quick verification of our mathematics so far, by returning to Eq.1 and Eq.2. By inserting our determined TON particle mass and charge from Eq.12 and Eq.13, we can calculate the force-distance product due to electrostatic attraction, F_E, and gravitation, F_G, which must be the same for an equilibrium condition at any distance, r. Therefore, from a re-arrangement of Eq.1:

$$F_G r^2 = Gm_T^2 = 6.674x10^{11}(3.3399x10^{-82})^2$$

$$= 7.4449x10^{-174} \qquad [Eq.14]$$

Similarly, from a re-arrangement of Eq.2:

$$F_E r^2 = Keq_T{}^2 = 8.897x10^9(2.8782x10^{-92})^2$$

$$= 7.4449x10^{-174} \qquad [Eq.\,15]$$

Applying the same rationale and equations discussed above to a conventionally-defined electron instead of a proton yields different results. While the mass of an electron is approximately $1/1836$ of a proton's mass, its electrical charge is identical to a proton's charge. Another complicating factor is that there is a large disparity in the scientific community regarding the size (diameter, radius) of an electron. Below, we list the values we used for the proton's parameters in the calculations presented above, and those for an electron.

Table 1. Properties of Classic Subatomic Particles

Parameter	Proton	Electron	Units
Mass	$1.672x10^{-27}$	$9.108x10^{-31}$	kg
Charge	$1.609x10^{-19}$	$1.609x10^{-19}$	Coulombs
Diameter	$1.75x10^{-15}$	$2.82x10^{-15}$ *	meters

*This is the "classic" value but estimates range from 10^{-13} through 10^{-18}

Using these values in Eq.5 and in Eq.10 through Eq.13 yields the following results, which we compare to those for the proton below:

Table 2.

	Result	Proton	Electron	Unit
Eq. 5	Charge Multiplier (CM)	$1.111x10^{18}$	$2.041x10^{18}$	(none)
Eq. 10	Diameter of a TON (d_T)	$8.261X10^{-34}$	$1.447x10^{-36}$	meters
Eq. 11	Number of TONs (n)*	$5.008X10^{54}$	$3.101x10^{64}$	(none)
Eq. 12	mass of a TON (m_T)	$3.339X10^{-82}$	$2.937x10^{-94}$	Kg
Eq. 13	Charge of a TON (q_T)	$2.878X10^{-92}$	$2.531x10^{-105}$	Coulombs

*In a proton or an electron

The differences in these quantities between proton-TONs (a collection of TON particles) and electron-TONs (number of TONs in the total number of photons) are not unexpected, especially since TON Particle Theory defines a photon shell surrounding an atom's nucleus instead of the conventional notion of an electron "particle." The photon shell contains a *large* number of captured photons (pairs of TON particle objects), and some photons may have greater mass than others, as discussed beginning in Chapter 3, and in Proofs 2 and 3. Hence, some of the calculated TON properties for an "electron" in the table above are only very coarse estimates, and not necessarily true within TON Particle Theory. The charge multiplier quantity is reasonable, but the diameter, number of TON particles, and charge of a TON particle are really indeterminate. Here, the diameter and number of TON particles assume a spherical electron particle volume and not a volume due to a hollow shell of some depth (thickness). Since we do not know the distance between the nucleus and the photon shell or the thickness of that shell, we cannot explicitly determine the number of TON particles, their diameter, or the charge of an individual TON particle. We also have the issue of the atomic state at which we make the measurements. An atomic element in a solid state will give us different values than those for the same element in a liquid or gaseous state. Our estimations are probably "close" to a true measurement, and fully investigating and better quantifying these TON particle parameters would be another very interesting research project for scientists.

The basic idea from these derivations is not to reconcile all differences, but to show a valid, rational, and plausible mathematical explanation of how like-charged TON particles can exist in close proximity to one another within the volume of a proton or of a spherical shell in a state of equilibrium between gravitational and electrostatic forces. From this equilibrium point, TON particles can exhibit the oscillation behavior described in the book and other Proofs. Clearly, to achieve this equilibrium state, TON particles must occur in significant numbers and extremely small sizes, and that is our premise. The reader may wish to experiment with the equations presented in this derivation, and see how different assumptions affect TON particle parameters.

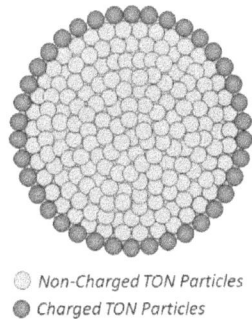

⊙ *Non-Charged TON Particles*
● *Charged TON Particles*

Figure P1-2. A Proton Composed of Charged TON Particles (Surface) and Non-Charged TON Particles

Charged Surface. To explore this possibility, we look at the outermost layer of like-charged TON particles that surround a proton, instead of considering the total number of TON particles filling its whole volume. As depicted in Figure P1-2, TON particles under the surface layer would have their charge absorbed (conducted) by the neighboring TONs due to their electrical permittivity. As we described in Chapter 2, the fusion process nullifies the repulsive force between touching TON particles, leaving gravity to hold those particles tightly together.

As in our charged volume derivations, we'll look at proton conditions and then apply the principles to classic electrons.

First, let's look at how TON particles would fit together in a single, one-TON deep layer on the surface of some sphere (a proton in this case) which is vastly larger than TON particles. We apply a packing factor, F_P, as we did for the charged-volume analysis. If we assume a square, two-dimensional lattice arrangement, the packing factor is simply the cross-sectional area of a spherical TON particle object divided by the area of a square whose dimensions are the diameter of a TON particle (d_T). In other words, the packing factor is the area of a circle divided by the area of a square whose four sides are tangential to the circle. This is absolutely correct for a flat surface. Since TON particles in the surface shell are significantly smaller than the proton sphere of diameter, d_P, upon which they lie, this is a reasonable approximation, as:

$$F_P = \frac{\pi \left(\frac{d_T}{2}\right)^2}{d_T^2} = \frac{\pi d_T^2}{4 d_T^2} = \frac{\pi}{4} = 0.7854 \qquad [Eq.\,16]$$

Where: $\pi(d_T/2)^2$ is TON particle cross sectional area, and
d_T^2 is the area of the tangential square around the
cross-sectional area

Next, we can consider that the two-dimensional "footprint" of each TON upon the surface area of a proton is its cross-sectional area. Hence, the number of TON particles in the surface shell, n, is the surface area of the proton divided by TON particle's cross-sectional area, and:

$$n = \frac{F_P \pi d_p^2}{\pi(d_T^2/4)} = \frac{(\pi/4)\pi d_p^2}{\pi(d_T^2/4)} = \frac{\pi d_p^2}{d_T^2} \qquad [Eq.\,17]$$

Where: $\pi d_P{}^2$ is the surface area of a proton sphere

When we multiply the number of TON particles in the shell by the surface area of a TON particle, we obtain a_s, the total surface area of TON particles in the shell:

$$a_S = \frac{\pi d_T^2 \pi d_p^2}{d_T^2} = \pi^2 d_P^2 \qquad [Eq.\,18]$$

Here, it may seem counter-intuitive that the total surface area of some number of TON particles is independent of TON particle diameter. This is best explained by a simple, flat-surface example. TON particles are so small compared to the size of a proton that surface curvature is negligible for a locally-small number of TON particles on the proton surface, and a flat surface approximation is valid. Let's look at a couple of situations. Say we have a 2-cm by 2-cm surface area and we place four 1-cm diameter spheres upon it as shown in Figure P1-3(a), or we place sixteen ½-cm diameter spheres upon it as shown in Figure P1-3(b). In either situation, the spheres cover the 2-cm by 2-cm surface area.

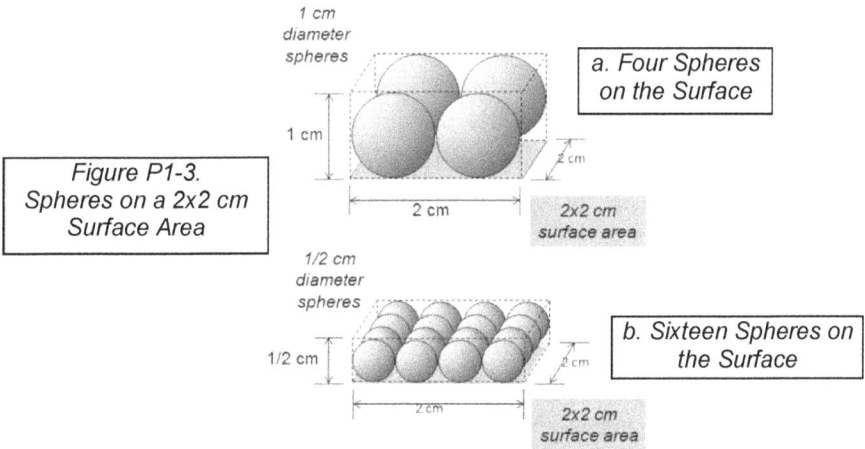

Figure P1-3. Spheres on a 2x2 cm Surface Area

1 cm diameter spheres

1 cm

2 cm

2 cm

2x2 cm surface area

a. Four Spheres on the Surface

1/2 cm diameter spheres

1/2 cm

2 cm

2 cm

2x2 cm surface area

b. Sixteen Spheres on the Surface

For the four-sphere situation, the total surface area of the four spheres, $a_{S(4)}$, is:

$$a_{S(4)} = 4\pi(1)^2 = 4\pi \ square \ centimeters$$

And the total volume, $v_{S(4)}$ of the four spheres is:

$$v_{S(4)} = \frac{4\pi(1)^3}{6} = \frac{4\pi}{6} = \frac{2\pi}{3} \ square \ centimeters$$

For the sixteen-sphere situation, the total surface area of the sixteen spheres, $a_{S(16)}$, is:

$$a_{S(16)} = 16\pi(1/2)^2 = 4\pi \ square \ centimeters$$

This is the same as the total surface area for the 4-sphere case above. The total volume, $v_{S(16)}$ of the sixteen spheres is:

$$v_{S(16)} = \frac{16\pi(1/2)^3}{6} = \frac{16\pi}{48} = \frac{\pi}{3} \ square \ centimeters$$

The effect is that when we fill an area with spheres, whether we use small spheres or large ones, the total surface area of all the spheres is the same. However, the total volume of the spheres changes, since the surface area remains constant while the diameters of the spheres change. Hence, as we decrease the diameter of the spheres, the total spherical surface remains constant and the total spherical volume decreases, so the surface area-to-volume ratio becomes larger and larger with smaller sphere diameters.

This gets us back to our charged surface derivation. If we multiply the expression for the number of TON particles in the shell, from Eq.17 by the volume of a TON particle, we obtain an expression for total TON particle volume in the shell:

$$v_S = \frac{\pi d_P^2}{d_T^2} \left(\frac{\pi d_T^3}{6} \right) = \frac{d_P^2 \pi^2 d_T}{6} \ square \ centimeters \qquad [Eq.19]$$

Recall the charge multiplier (CM) originally calculated by Eq.4 and Eq.5, and used to calculate TON particle diameter in Eq.9 and Eq.10 for the charged volume situation. In that case, the charge multiplier was attributed to the proton-to-TON particle surface area ratio as a function of TON particle diameter. This was because the accumulated TON particle surface area was a function of TON particle diameter, while proton and TON particle-equivalent volumes were constant for changes in TON particle diameter. For the charged shell situation, the charge multiplier is attributed to the proton-to-TON particle accumulated shell volume ratio, because accumulated TON particle shell volume changes as a function of TON particle diameter while accumulated TON particle surface area remains constant.

Hence, by dividing the volume of a proton by this expression for total shell volume, we obtain an expression for the charge multiplier (CM), as:

$$CM = \frac{\left(\frac{\pi d_P^3}{6}\right)}{\frac{d_P^2 \pi^2 d_T}{6}} = \frac{d_P}{\pi d_T} \qquad [Eq.\,20]$$

And re-arranging terms gives us TON particle diameter, as:

$$d_T = \frac{d_P}{\pi CM} = \frac{(1.7536x10^{-15})}{\pi(1.111x10^{18)}} = 5.0222x10^{-34}\ m \qquad [Eq.\,21]$$

This result is in very good agreement with TON particle diameter value of $8.2612x10^{-34}$ m which we calculated in Eq.10, for the charged-volume case. The difference is generally due to the different geometric interpretations of the two possibilities. In assessing both the charged volume and charged surface cases, we made some mathematically-simplifying assumptions regarding a packing factor, or how closely TON particles are spaced in elemental spheres or surface shell volumes. If we use a slightly looser or less dense packing factor of 0.3183 instead of 0.5236 for the charged volume case, Eq.10 gives us a TON diameter of 5.022^{-34} meters, essentially the same as from Eq.21 for the charged surface case.

Now we can use Eq.17 to calculate the number of TON particles in the surface shell, n_s, as:

$$n_S = \frac{\pi d_p^2}{d_T^2} = \frac{\pi(1.7536x10^{-15})^2}{(5.02x10^{-34})^2} = 3.8302x10^{37} \qquad [from\ Eq.\,17]$$

The number of TON particles in a whole proton can be obtained as before from Eq.11. Here we use the volumetric packing factor of

0.5236 we calculated earlier for the whole proton volume, and obtain:

$$n = 0.5236\frac{d_P^3}{d_T^3} = 0.5236\frac{(1.7536x10^{-15})^3}{(5.02x10^{-34})^3} = 2.2290x10^{55} \qquad [from\ Eq.\ 11]$$

We can calculate a TON particle mass and TON particle charge by re-using Eq.12 and Eq.13. The mass of a TON particle is the mass of a proton divided by the number of TON particles in the proton, as:

$$m_T = \frac{1.6726x10^{-27}}{2.2290x10^{55}} = 7.5039x10^{-83}kg \qquad [from\ Eq.\ 12]$$

And the charge of a TON particle is the charge of a proton divided by the charge multiplier from Eq.5 to get the proton-equivalent TON particle charge, and then divided by the number of TON particles carrying a charge, or the number of TON particles in the outer layer, as:

$$q_T = \frac{q_P}{(CM)n} = \frac{1.6026x10^{-19}}{(1.111x10^{18})(3.8302x10^{37})}$$

$$= 3.7632x10^{-75}Coulombs \qquad [from\ Eq.\ 13]$$

We can again do an assessment of the charged-layer mathematics, by returning to Eq.1 and Eq.2. By inserting our determined TON particle mass and charge from the layer, we can calculate the force-distance product due to electrostatic repulsion, F_E, and gravitation, F_G, which would be the same for an equilibrium condition at any distance, r. Similar to the charged volume case, from a re-arrangement of Eq.1 we obtain:

$$F_G r^2 = Gm_T^2 = 6.674x10^{11}(7.5039x10^{-83})^2$$

$$= 3.7581x10^{-175} \qquad [from\ Eq.\ 14]$$

Similarly, from a re-arrangement of Eq.2 we get:

$$F_E r^2 = Keq_T^2 = 8.897x10^9(3.7632x10^{-75})^2$$

$$= 1.2727x10^{-139} \qquad [from\ Eq.\ 15]$$

In this case, unlike the charged volume case, electrostatic and gravitational forces do not match, and we do not achieve equilibrium amongst TON particles in the charged layer. In fact, the electrostatic repulsive force, F_E, dominates the gravitational attractive force, F_G, at any distance, r, within the surface layer. However, we have yet to consider the attractive force of the mass of non-charged TON particles comprising the volume of the proton sphere not contained in the 1-TON particle deep surface layer. Figure P1-4 portrays these forces graphically, but the force vectors (arrows) are not to scale.

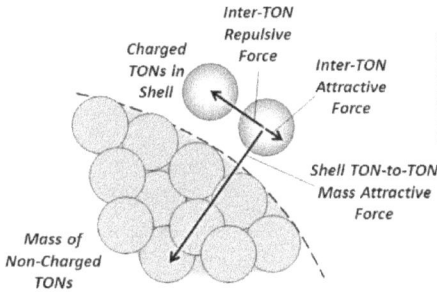

Figure P1-4. Forces at Work in the Charged-Surface (Shell) Case

In this situation, as the figure illustrates, the surface shell TON particles are not only actively repulsing each other, but are attracted to the mass of aggregated non-charged TON particles beneath the surface. Going back to Coulomb's and Newton's laws, we can calculate these forces and assess their impacts for a nominal, un-quantified distance, r.

To begin, the attractive gravitational force between two charged TON particles in the outer layer is:

$$F_G = \frac{Gm_T m_T}{r^2} = \frac{Gqm_T^2}{r^2} = \frac{3.7581x10^{-175}}{r^2} \ newtons \ \ [from \ Eq. 1]$$

The repulsive gravitational force between two charged TON particles in the outer layer is:

$$F_E = \frac{Keq_T q_T}{r^2} = \frac{Keq_T^2}{r^2} = \frac{1.2727x10^{-139}}{r^2} \ newtons \ \ [from \ Eq. 2]$$

The attractive gravitational force between a TON particle in the outer proton layer to the mass of the aggregated TON particles beneath the surface is:

$$F_{MT} = \frac{Gm_T m_{P-S}}{r^2} = \frac{8.3766x10^{-120}}{r^2} \ newtons \ \ [from \ Eq. 1]$$

Here, we use the quantity $m_{P\text{-}S}$ to represent the mass of the aggregated TON particles in the proton minus the mass of TON particles in the surface layer. As these equations show, electrostatic repulsion force is concentrated *in the surface layer.* The fusion process reduces the electrostatic repulsive force within the proton, allowing gravity to maintain the adhesion between individual TON particles (Chapter 2). The surface layer TON particles retain a repulsive force along their outer surfaces, while the repulsive force between that layer and the TON particle layer below is partially canceled by the permittivity of the TON particles themselves.

Proof 2: TON Particles in Oscillation

In Proof 1 we established that both electrostatic (repulsive) and gravitational (attractive) forces affect the position of TON particles in free space. When two TON particles are in close proximity to one another, then Coulomb's electrostatic (repulsive) force, being the stronger of the two forces, would tend to push them apart. We have established a TON particle's diameter to be nominally between 10^{-34} and 10^{-36} meters, with a mass estimated to be between 10^{-82} and 10^{-94} kilograms. Needless to say, these are extremely small particles.

In several chapters, we describe how a pair of TON particles linked in an entanglement (a photon) in free space move back-and-forth (vibrate) in an attempt to balance the two forces. In Proof 2, we expand these discussions to show how two TON particles of similar charge and mass may enter into a steady oscillation or vibratory state. We will further show that the mass and charge of those two objects determine their specific vibrational frequency. As with our other proofs, we pursue the mathematics to the extent of plausibility, and forgo highly complex, deterministic modeling which is a bit beyond the scope of this work.

Let's look at some basic equations that describe oscillatory motion. Our handy physics book[71] defines the displacement, or distance traveled for some particle as a function of time as:

$$x = A\cos(\omega t + \delta) \qquad [Eq. 1]$$

Where: x = Displacement (meters)
A = Displacement constant, or maximum range in a direction
ω = Angular frequency of the motion (radians per second)
δ = Phase shift (radians per second)
t = Time (seconds)

And the angular frequency may be further broken down into subcomponents as:

---

[71] Halliday, D. and R. Resnick. *Physics for Students of Science and Engineering.* Wiley and Sons. 1965.

$$\omega = \sqrt{\frac{k}{m}} = 2\pi f_0 \qquad [Eq.\,2]$$

Where: ω = Angular frequency of the motion (radians per second)

k = Force constant of the elastic member of the system (N/m) (analogous to a spring constant for a mechanical system)

m = Mass (kg)

f_O = Frequency of oscillation (Hz)

We can also re-state Eq.2 to provide an expression for oscillation frequency, f_O, as:

$$f_0 = \left(\frac{1}{2\pi}\right)\sqrt{\frac{k}{m}} \qquad [Eq.\,3]$$

The k/m term is a measure of the ability of the two masses to move as a function of electrostatic repulsion. If we were dealing with a spring, electrostatic force would represent the stiffness of the spring or the spring constant. In our case, the two TON particles will exhibit periodic motion due to the mass of each particle attracting the other by gravity, compressing our "spring" until electrostatic force can overcome the mass momentum and propel both masses away from each other. As k approaches zero, so does ω, and the system remains at its initial displacement with no oscillation.

The equations here describe the relative motion of each TON particle to each other as an entangled pair. Both TON particles will be moving, as the force acts equally on both. In three dimensions, more complex motions are possible, but for our example, a simple, single-vector case, will demonstrate the plausibility with minimal complications.

Given this expression of displacement in Eq.1, we can derive equations for velocity and acceleration simply by differentiating the displacement equation, which gives us:

$$v = \frac{dx}{dt} = -A\omega \sin(\omega t + \delta) \qquad [Eq.\,4]$$

$$a = \frac{d^2 x}{dt^2} = -A\omega^2 \cos(\omega t + \delta) \qquad [Eq.\,5]$$

Where: v = Velocity (m/sec)
   a = Acceleration (m/sec2)

Now let's enter some of the values from Proof 1 into the displacement and velocity equations and see what happens. We'll start with some initial, nominal values for displacement constant (A), mass (m), and force constant (k):

Let A = 10^{-33} m, or about an order of magnitude larger than our estimated TON diameter calculated in Proof 1. Recall that in Proof 1 we assumed a tight packing factor, so we are defining our range of displacement to be greater than a TON particle radius.

Let m = 10^{-82} kg, or about the mass we estimated for a TON particle in Proof 1, and

Let k = 1.0, since we need a starting point for our calculations.

Using these values, Eq.3 says that:

$$f_0 = \left(\frac{1}{2\pi}\right)\sqrt{\frac{1}{10^{-82}}} \ = 1.59 \times 10^{40}\ Hz \quad [from\ Eq.\,3]$$

Then, by inserting these same nominal values into Eq.4 and Eq.5 we can observe the oscillatory behavior, as shown in Figure P2-1. Here, the maximum displacement is simply the displacement constant, A = 10^{-33} m, which we chose, and the maximum velocity is the value, v(max) = $A\omega$ = 1x10^8 m/sec, from Eq. 4. These are sensible results, as the maximum velocity does not exceed the speed of light (3x10^8 m/sec), and the maximum displacement is larger than the estimated TON diameter because we defined it that way. This suggests that two TON-sized particles can form oscillating pairs whose frequency is on the order of 10^{40} Hz. Hence, oscillation is indeed plausible with TON particle size masses in the range we estimated in Proof 1.

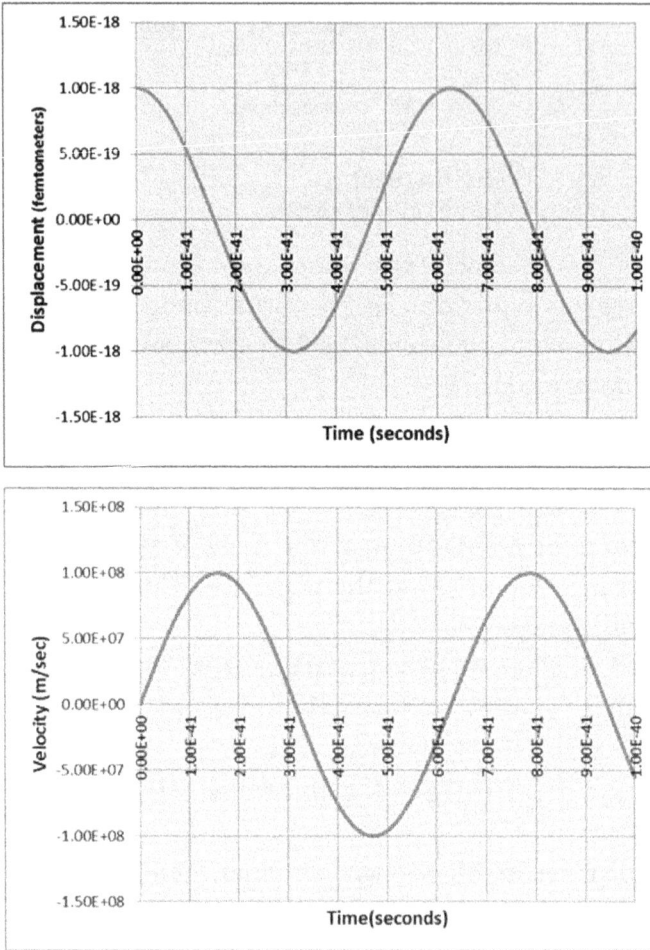

Figure P2-1. Displacement (top) and Velocity (bottom) Characteristic for Initial Oscillation Values

From the frequency and velocity equations, it is apparent that more massive TON particle objects (as a stellar fusion process would create (Chapter 2)) will oscillate at lower frequencies. When we plot frequency as a function of mass, with different force constant (k) values, as depicted in Figure P2-2, we clearly see that larger masses induce lower oscillation frequencies, and we get a family of curves for different k values. This force constant is arbitrary at this point, as we cannot really define the elasticity or "spring constant" for such small particles, which we have not experimentally measured yet. It's not something you can simply measure with a torsion meter.

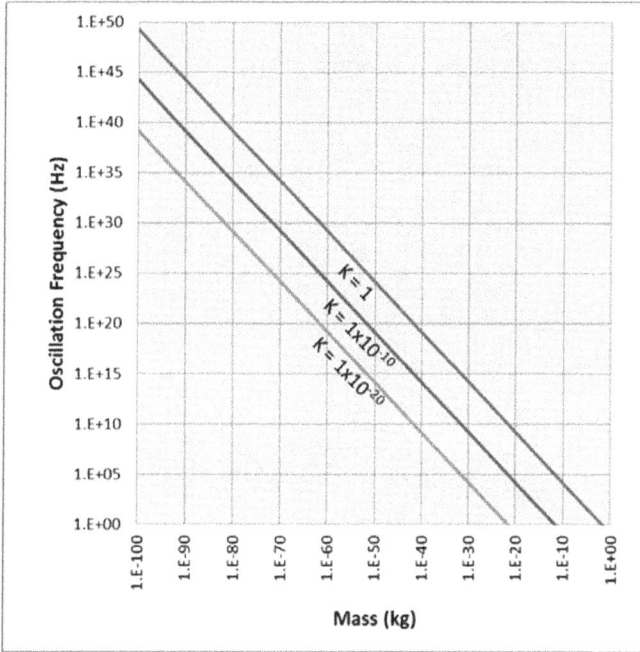

Figure P2-2. Mass, Force Constant, and Oscillation Frequency Relationships

Summarizing, we have shown the *plausibility* of TON particle pairs oscillating at very high frequencies, on the order of 10^{30} Hz to 10^{40} Hz for our estimated TON particle mass, depending upon how we set our force constant. We have applied sufficient mathematical analysis, based on known laws of physics, to arrive at this level of plausibility estimation. Some day in the future it would be interesting to see how results from more deterministic simulation or maybe even measurement of these effects compare with these initial calculations.[72]

---

[72] Halliday, D. and R. Resnick. *Physics for Students of Science and Engineering.* Wiley and Sons. 1965

Proof 3: Magnetic and Electrostatic Balance in the Photon Shell

Proof 3 determines the plausibility of captured photons forming a stable photon shell around an atomic nucleus to create a self-supporting structure to replace the classic model of an electron orbiting said nucleus. Using the ideas presented in Chapters 3 and 4, we will show how we can achieve a balance of forces between the atomic nucleus and the photon shell.

For example, using the hydrogen atom as our basis, classic physics has a notional view of hydrogen composed of one proton and one orbiting electron. In Chapter 6, we introduced a new atomic model with the proton surrounded by a spherical photon shell, which is one-photon thick.

We begin with the spherical geometry with the proton surrounded by the photon shell. Using the classic measured diameter of a proton (1.75×10^{-15}), and the measured diameter of the hydrogen atom at Bohr's limit (10^{-10}), we find that the proton is significantly smaller. The measured diameter of a proton is several orders of magnitude larger than the distance between the shell and the proton. Also, individual TON particles and photons are many orders of magnitude smaller than a proton. It is this disparity in size and distance that allows TON Particle Theory atomic model to achieve force balance. To discuss the mathematics involved, we will use a disproportionate scale in our diagrams to portray the particle geometry clearly.

Case 1: Photon-Proton Relationship

We begin with the unit photon, which would be two TON particles within the photon entanglement. When we assemble a photon shell equivalent in mass to the classic notion of an electron particle, we discovered that the photons can no longer vibrate in this configuration. Further research into measured electron behavior revealed that the captured photons would rotate around a vector axis perpendicular to the center of a proton (nucleus) through the photon's center of rotation, as Figure P3-1 illustrates.

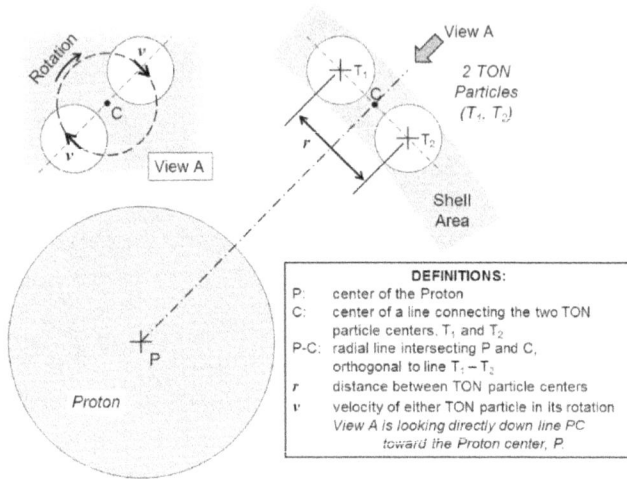

Figure P3-1. Proton and Photon Shell in a 3-Axis Spinning Dynamic

Note that the captured photons in this model are *not* orbiting the proton/nucleus. It is these spinning TON particles that scientists measured and mistakenly determined to be electron particle rotation or spin. Only spinning TON particles would generate the electromagnetic signature needed for them to measure "electron spin."

Coulomb's law of electrostatic repulsion, as mentioned in Proof 1, states that two objects with identical charges, *q*, at a distance or range, *r*, results in repulsive force F_E:

$$F_E = \frac{K_e q^2}{r^2} \qquad [Eq. 1]$$

Where: F_E = Electrostatic Force, in Newtons (N)
   q = Electrical charge on each TON particle (identical), in Coulombs (C)
   r = Range or distance between TON particles, in meters (m)
   Ke = Coulomb's electrical attraction constant = 8.987x10⁹ (N-m2/C2)

The formula for magnetic attraction or repulsion between two charges is somewhat similar to Eq.1, as:

$$F_M = \frac{\mu_0 q^2 v^2}{4\pi r^2} \qquad [Eq.\,2]$$

Where: F_M = Magnetic Force, in Newtons (N)
q = Electrical charge on each TON particle (identical), in Coulombs (C)
r = Range or distance between TON particles, in meters (m)
v = The velocity at which the particles (TON particles) are moving
μ_0 = The permeability of free space, a constant = $4\pi \times 10^{-7}$ (Henries/m)

At this point we'll set aside the gravitational forces and concentrate solely upon the electrostatic and magnetic forces defined above. We can do this because the distances between the proton and the photon shell results in a negligible gravitational effect, compared to the electrostatic and magnetic forces. Our purpose in this proof is to show how the electrostatic force, which is repulsive, may be balanced or overcome by an equal or greater magnetic force, which is attractive, caused by TON particles spinning.

To establish a force equilibrium condition between the two spinning TON particles within each photon, we use a mathematical convention where we express attractive forces as positive and repulsive forces as negative, as in:

$$F_M - F_E = 0, \quad or \quad F_M = F_E \qquad [Eq.\,3]$$

Substituting the quantities of Eq.1 and Eq.2 into Eq.3 gives us:

$$\frac{\mu_0 q^2 v^2}{4\pi r^2} = \frac{K_e q^2}{r^2}$$

Cancelling like terms on both sides of this equation reduces it to:

$$\frac{\mu_0 v^2}{4\pi} = K_e \qquad [Eq.\,4]$$

Now, re-arranging terms gives us an expression for the velocity, v, required to achieve equilibrium in terms of the other remaining quantities, which are all physical constants, as:

$$v^2 = \frac{4\pi K_e}{\mu_0} \qquad [Eq.5]$$

However, Coulombs constant, K_e is defined in terms of electrical permittivity as:

$$K_e = \frac{1}{4\pi \varepsilon_0} \qquad [Eq.6]$$

Where: ε_0 = The electrical permittivity of free space (Farads/m)

So, we can now substitute the quantities of Eq.6 for K_e in Eq.5, which gives us:

$$v^2 = \frac{4\pi}{\mu_0 4\pi \varepsilon_0} = \frac{1}{\mu_0 \varepsilon_0} \qquad or$$

$$v = \frac{1}{\sqrt{\mu_0 \varepsilon_0}} \qquad [Eq.7]$$

Here, we introduce another physical constant, the speed of light, c, which is defined in terms of permeability and permittivity of free space as:

$$c = \frac{1}{\sqrt{\mu_0 \varepsilon_0}} \qquad [Eq.8]$$

Since Eq.7 and Eq.8 are identical, $v = c$, *the velocity required for equilibrium is the speed of light.* This last fact is crucial, because we know that the two TON particles inside the photon were oscillating as a function of the speed of light before they were captured. Therefore, it is essential that the vibrational momentum of the particles be translated into angular momentum, which supports our findings that they must spin in the photon shell after capture.

Further, all the TON particle pairs within all the captured photons in the photon shell must rotate in the same direction. Otherwise, adjacent captured photons would cancel out each other's magnetic field and disrupt the photon shell integrity. Luckily, as the approaching photon is absorbed into the photon shell, the existing spin of the already captured photons will determine in which direction the two new TON particles will spin.

Next, let's consider the effect of the proton upon the captured photons, by first looking at the forces in effect between the proton and the two TON particles within each captured photon. Again, electrostatic repulsion dominates gravitational attraction, and we are interested in seeing how magnetic and electrostatic forces may reach a state of equilibrium, thus allowing the gravitational forces to "bind" the photon shell around the proton (nucleus).

We look into two possibilities here, and a third possibility later in Proof 4. One possibility here is that the photon shell and/or the proton are spinning (have a complex, 3-axis, rotational velocity dynamic). The other case is that they are both more or less stationary. With rotational velocity, magnetic forces apply. Without it, there would be no appreciable magnetic effect, and we would need some other means to counter the repulsive electrostatic force. In either case, TON particles within the photons are rotating on their local axes, thus creating a magnetic field.

Case 2: Rotational Velocity.

Take the case where the photon shell and the proton are rotating in different directions. The electrostatic and electromagnetic force relationships here are similar to those of Eq.1 and Eq.2 for the photon, but the charges, velocities, and distances are different, as Figure P3-2 shows. The proton's charge is evenly distributed over the outermost layer of the proton surface, as we explained in Proof 1 for the charged surface possibility. Both the proton and the photon shell are engaged in an undetermined, 3-axis, spinning dynamic at some spin velocity, v, as shown.

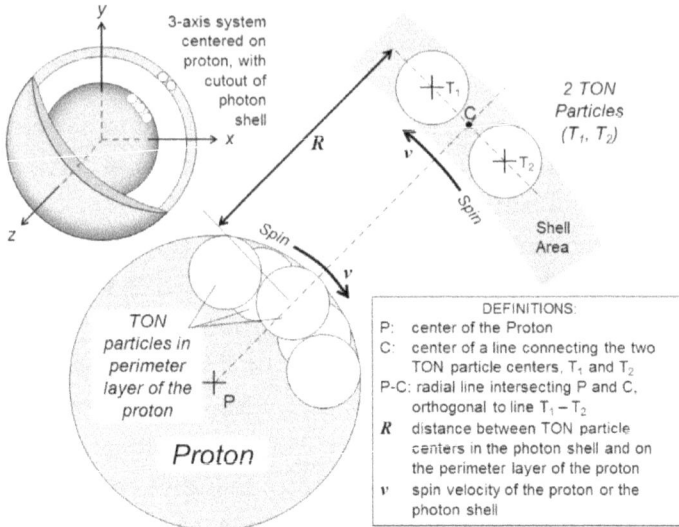

Figure P3-2. Proton and Photon Shell in a 3-Axis Spinning Dynamic

In Case 2, unlike the photon orientation situation discussed in Case 1, we have different charges for TON particles in the photon shell and TON particles in the surface layer of the proton. Coulomb's law for this case defines the electrostatic force of repulsion (since TON particles are similarly charged), F_E, as:

$$F_E = -\frac{K_e q_T q_P}{R^2} \qquad [Eq.\,9]$$

Where: F_E = Electrostatic Force, in Newtons (N)
   q_T = Electrical charge on a shell TON, in Coulombs (C)
   q_P = Electrical charge on a proton surface TON, in Coulombs (C)
   R = Distance between shell TON particles and proton TON particles, in meters (m)
   Ke = Coulomb's electrical attraction constant
     = 8.987x109 (N-m2/C2)

The formula for magnetic attraction or repulsion between two charges is similar to Eq.2, as:

$$F_M = \frac{\mu_0 q_T q_P v^2}{4\pi R^2} \qquad [Eq.\,10]$$

Where: F_M = Magnetic Force, in Newtons (N)

q_T = Electrical charge on a TON particle within a photon, in Coulombs (C)

q_P = Electrical charge on a proton surface TON, in Coulombs (C)

R = Distance between shell TON particles and proton surface TON particles, in meters (m)

v = The spin velocity of the proton or shell

μ_0 = The permeability of free space, a constant = $4\pi \times 10^{-7}$ (Henries/m)

At this point, substituting the quantities of Eq.9 and Eq.10 into our earlier force-balancing equation, Eq.3 gives us a similar relationship, the only difference being the charge (q) and distance (R) quantities:

$$\frac{\mu_0 q_T q_P v^2}{4\pi R^2} = \frac{K_e q_T q_P}{R^2}$$

And since these charge and distance quantities cancel out on both sides of the equation, we get a result identical to that of Eq.4, as:

$$\frac{\mu_0 v^2}{4\pi} = K_e \qquad [Eq.\,11, same\ as\ Eq.\,4]$$

At this point the rest of the derivation is identical to that for the rotating photon situation, from Eq.4 through Eq.8. As before, $v = c$ and the spinning velocity on the shell, v, and the proton velocity, again v, required for equilibrium are both the speed of light. So, the proton and its surrounding photon shell must be spinning in some manner at the speed of light to achieve equilibrium. The exact spin patterns, over the three-dimensional axes of the proton (nucleus) and the photon shell are a matter of speculation and somewhat too mathematically complex to be addressed here. Our aim is simply to pose a plausible force balancing solution.

Case 3: No Rotational Velocity Case.

Consider the case where neither the photon shell nor the proton has a rotational velocity. We apply some numbers to our earlier equations to assess force balancing plausibility. Figure P3-3 shows the attributes and notations for this case. This case examines how the photon shell maintains its integrity with a significant electrostatic repulsive force, F_E, between TON particles in the proton and in the photon shell. There are of course gravitational forces between the two TON particles in each photon in the photon shell and between each captured photon and the proton (not shown, and insignificant compared to the electrostatic forces).

Once again, the repulsive force, F_E, is actually a vector quantity, which has a tangential component, $F_{E,T}$, trying to push the two TON particles apart inside the captured photon. Meanwhile, the primary vector component of F_E is pushing outward by the same amount on all photons, so there is no force differential in that direction trying to push any photon away from any nearby photon, in the radial direction. So, only the tangential force component, F_E, needs to be balanced to maintain photon shell integrity.

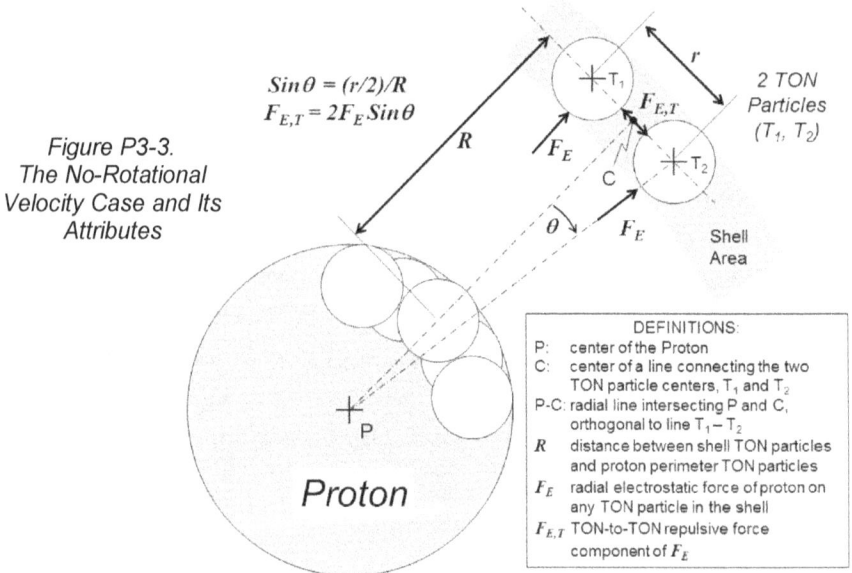

Figure P3-3.
The No-Rotational Velocity Case and Its Attributes

$$Sin\,\theta = (r/2)/R$$
$$F_{E,T} = 2F_E\,Sin\,\theta$$

2 TON Particles (T_1, T_2)

DEFINITIONS:
P: center of the Proton
C: center of a line connecting the two TON particle centers, T_1 and T_2
P-C: radial line intersecting P and C, orthogonal to line $T_1 - T_2$
R distance between shell TON particles and proton perimeter TON particles
F_E radial electrostatic force of proton on any TON particle in the shell
$F_{E,T}$ TON-to-TON repulsive force component of F_E

Now, let's calculate the number of TON particles in the photon shell. Using the same diameter we calculated for TON particles in the proton (Proof 1), the area of a circular TON "footprint" on the surface of a sphere of radius = R, divided into the surface area of that sphere represents an approximation of the number of TON particles in the photon shell at that same distance, R, from the proton, when multiplied by a packing factor, F_P. We'll use a packing factor of $\pi/4$ (value), which is the area of a circle divided by the area of a square whose sides are the radius of the circle – a reasonable assumption since the size of TON particles are many orders of magnitude smaller than the size of the photon shell sphere. Then, the number of TON particles in the photon shell is approximately:

$$N_T = \frac{F_p 4\pi R^2}{\pi (r/2)^2} = \frac{F_p 16R^2}{r^2} = 1.394x10^{47} \qquad [Eq.\,12]$$

Where: N_T = Number of TON particles in the photon shell
R = Radius of the photon shell sphere (Bohr's distance = 5.29x10^{-11} m)
d_T = Diameter of a TON particle (5.022x10^{-34} m)
F_P = $\pi/4$

Now, we divide the known mass and charge of a classic electron particle by the number of TON particles, to get the mass and charge of a TON particle in the photon shell, as:

$$m_{T,B} = \frac{m_E}{N_T} = 6.531x10^{-78}\,kg \qquad [Eq.\,13]$$

$$q_{T,B} = \frac{q_E}{N_T} = 1.149x10^{-66}\,C \qquad [Eq.\,14]$$

Where: N_T = Number of TON particles in the photon shell
$m_{T,B}$ = Mass of a TON in the photon shell
$q_{T,B}$ = Charge of a TON in the photon shell
m_E = mass of an electron (9.107x10-31 m)
q_E = charge of an electron (1.602x10-19 m)

Next, we can calculate the radial electrostatic force acting upon a TON due to the proton's charge, from Coulomb's Law, as:

$$F_{E,R} = -\frac{K_e q_P q_T}{R^2} = 5.911x10^{-55} \qquad [Eq.\,15]$$

Where: $F_{E,R}$ = Radial electrostatic Force, in Newtons (N)
  q_T = Electrical charge on a TON Particle inside a
       photon shell ($1.149x10^{-66}$ C)
  q_P = Electrical charge on the proton, in Coulombs
       ($1.602x10^{-19}$ C)
  R = Distance between photon shell TON particles
       and proton surface TON particles ($5.29x10^{-11}$ m)
  K_e = Coulomb's electrical attraction constant
       = $8.987x10^9$ (N-m^2/C^2)

Now we can calculate the electrostatic repulsive force between adjacent photons in the photon shell due to the tangential component of the radial force, as:

$$F_{E,R,T} = -2F_{E,T}\,Sin\,\theta = 1.117x10^{-77}N \qquad [Eq.\,16]$$

Where: $F_{E,R,T}$ = Tangential part of radial electrostatic Force,
        in Newtons (N)
    F_E = Radial electrostatic Force ($5.911x10^{-55}$ N), and:
    $Sin\,\theta$ = r/R = $9.452x10^{-24}$

Recall that the magnetic force between TON particles in a photon shell due to their rotation was expressed in Eq.2. Now, for the quantities we are considering it becomes:

$$F_{M,T} = \frac{\mu_0 q^2 v^2}{4\pi r^2} = 1.186x10^{-56}N \qquad [Eq.\,17, from\,Eq.\,2]$$

Where: $F_{M,T}$ = Magnetic Force between photons, in Newtons (N)
  q = Electrical charge on each TON particle in the
       photon shell ($1.149x10^{-66}$ C)
  r = Range or distance between TON particles in the
       photon shell ($1x10^{-33}$ m)
  v = TON particle's velocity in rotation (speed of
       light = 299,792,458 m/s)
  μ_0 = The permeability of free space, a constant
       = $4\pi\,x10^{-7}$ (Henries/m)

However, as shown earlier in this proof, the tangential electrostatic and magnetic forces cancel each other out exactly when TON particle velocity in the photon shell rotates at the speed of light. In other words, $F_{E,T} = F_{M,T}$, so the total tangential force between photon shell photons, $F_{T,B}$, is:

$$F_{T,B} = F_{M,T} - F_{E,T} - 2F_{E,R,T} = F_{E,R,T} = 2.234x10^{-77} N \quad [Eq. 18]$$

Now, looking back at the magnetic force calculation, let's consider permeability. The magnetic force is directly proportional to the permeability, which we have been assuming is that of free space (or a vacuum), μ_0. But this is a matter of speculation. Real materials we can measure exhibit permeability values from $1.0000003\mu_0$ to $1000000\mu_0$, which is quite a range of variability.

We see from Eq.16 and Eq.17 that the tangential component of the radial electrostatic force we need to overcome, $F_{E,R,T}$, is a small fraction, F_F, of the tangential magnetic force between photon shell photons, as:

$$F_F = \frac{F_{E,R,T}}{F_{M,T}k} = 1.884x10^{-21} \; (dimensionless) \quad [Eq. 19]$$

Where: F_F is the tangential force fraction (dimensionless)
  $F_{M,T}$ = Magnetic Force between photon shell photons
    $(1.186 \, x10^{-56}$ N)
  $F_{E,R,T}$ = Tangential part of radial electrostatic force
    $(1.117 \, x10^{-77}$ N)

These results indicate than an increase in permeability, μ_0, of about two parts in 10^{21} would be sufficient for the magnetic force to balance both the photon shell electrostatic force and the tangential component of the radial electrostatic force. This change is a very small deviation from μ_0, and who can say what the intra-atomic permeability really is? We clearly need more research to establish the internal atomic permeability and permittivity.

There are still issues about the true number of photons in a minimal (or base) state hydrogen atom's photon shell[73]. TON Particle Theory clearly shows that the number of photons will differ depending upon the atom's physical state and the environment in which it exists.

---

[73] An atom at base state (zero degrees Kelvin) would contain only unit photons. As mass/energy is absorbed by the photon shell, the individual photons in that shell can be of any larger mass.

The theory also shows that the number is in constant flux as energy and mass change.

We used a fairly efficient packing factor of $\pi/2$ and a photon shell-to-proton distance equal to Bohr's distance as nominal values in this proof. A different packing factor would simply distribute the photon shell charge and mass over fewer photons. A closer distance between the proton and the photon shell would bring the photons almost into contact with the proton, a premise which Chapter 5 addresses about neutrons. Any combination of these two independent variables will still achieve the force balance we have discussed, and the reader may wish to experiment with the various equations herein to verify these conditions and/or to test his or her own derivative ideas in this domain.

To summarize Proof 3:

- We looked at atomic level magnetic and electrostatic forces, and how they could balance or cancel one another and allow the gravitational forces to bind the photon shell to the proton at the consensus-measured mass ratio.
- The overlying forces that control multiple photons in the photon shell and the mass of TON particles in the proton's outer layer, establish the conditions where the individual TON particle pairs in each photon must rotate in the same direction to avoid magnetic force cancellation.
- Scientists have already established that all electrons spin, and TON Particle Theory clearly shows how that spin is created for each atom.
- Our proof shows the *plausibility* of maintaining the gravitational entanglement of TON particles at the atomic level.

A more rigorous mathematical model will be needed as science verifies and validates TON Particle Theory and creates instruments capable of measuring the distances, velocities, and physical constants for this new atomic view. TON Particle Theory creates many useful projects for research organizations and university laboratories to resolve these issues.

Proof 4: More Concepts of Nucleus-to-Photon Shell Force Balance

Since the first publication of TON Particle Theory in *I Killed Schrö-dinger's Cat*, we conducted additional analyses to the force balance situations initially developed in Proof 3. We further addressed the balance of forces between an atom's nucleus and its photon shell, specifically where *TON particle pairs (photons) in the photon shell are in a rotational motion and TON particles in the nucleus are either motionless or at least moving at a very slow rate compared to those in the photon shell.*

To begin let us briefly summarize Newton's and Coulombs Laws, from Proof 1, just to refresh our understanding of these foundation principles.

Newton's law states that two masses m_1 and m_2 at a distance or range of r from one another exert a gravitational force of attraction, F_G, as:

$$F_G = \frac{Gm_1m_2}{r^2} \qquad [Eq.\,1]$$

Where: F_G = Gravitational Force, in Newtons (N)
         m_x = Mass, in kilograms (kg)
         r = Range or distance, in meters (m)
         G = Newton's gravitational constant
                = 6.674×10^{-11} (N-m^2/kg^2)

Coulomb's law states that two bodies with electrical charges q_1 and q_2 at a distance or range of *r* from one another exert an electrostatic force of attraction (if oppositely charged) or repulsion (if similarly charged), F_E, as:

$$F_E = \frac{Keq_1q_2}{r^2} \qquad [Eq.\,2]$$

Where: F_E = Electrical Force, in Newtons (N)
         q_x = Electrical charge, in Coulombs (C)
         r = Range or distance, in meters (m)
         K_e = Coulomb's electrical attraction constant
                = 8.987×10^9 (N-m^2/C^2)

To begin, we review the Properties of Classic Subatomic Particles (Table 1) and our estimated Properties of TON Particles (Table 2) from Proof 1f.

Table 1. Properties of Classic Subatomic Particles

Parameter	Proton	Electron	Units
Mass	1.672×10^{-27}	9.108×10^{-31}	kg
Charge	1.609×10^{-19}	1.609×10^{-19}	Coulombs
Diameter	1.75×10^{-15}	2.82×10^{-15} *	meters

*This is the "classic" value but estimates range from 10^{-13} through 10^{-18}

Table 2. Estimated Properties of TON Particles

Property	Value	Units
Diameter of a TON (d_T)	8.261×10^{-34}	meters
Number of TONs (n) in a Proton	5.008×10^{54}	(none)
Mass of a TON (m_T)	3.339×10^{-82}	kg
Charge of a TON (q_T)	2.878×10^{-92}	Coulombs
Proton Volume Packing Factor	0.5236	
Proton Surface Packing Factor	$0.7854 = \pi/4$	

Basic Force Balance. In Proof 1, we evaluated force balancing by setting the electrical and gravitational forces equal to one another, basically saying that Eq.1 = Eq.2, which gave us the following result:

$$F_G = F_G, \ so \ \frac{Gm_1m_2}{r^2} = \frac{Keq_1q_2}{r^2} \quad [Eq.\,3]$$

This result shows that the gravitational and electrical forces would mathematically "cancel out" of each side of the equation. Physically, this meant that the electrical and gravitational forces would balance at any range (set of all real r values), if the various constants, masses, and charges satisfied this identity:

$$Gm_1m_2 = Keq_1q_2 \quad [Eq.\,4]$$

Originally, we wanted (1) to derive mass and charge values for individual TON particles in close proximity and (2) to show that like-charged TON particles could exist in a balanced-force condition with one another. We proved this using the assumption of charge distribution over the surface of a proton and mass distribution, of course, over the volume of said proton. In Proof 1 we evaluated Eq. 4 under this assumption, using the same masses and charges for each of two TON particles in proximity, successfully proved this assumption, and derived the characteristic values for TON particles shown in Table 2 above. We can think of this prior work as describing the means by which TON particles may exist in close proximity at the subatomic level.

Macroscopic or Atomic Level Force Balance.

Now we confront a more *macroscopic situation at the atomic level*, where there is a nucleus comprised of a large number of individual TON particles and a surrounding photon shell also comprised of a large number (but not as large as in the nucleus) of TON particle pairs (photons). Here we also require force balance, basically to suspend each photon in the photon shell at some distance, D, from the center of the nucleus. If we can establish such a force balancing relationship between a spherical nucleus and a photon shell consisting of just two TON particles entangled (unit photon), and if D is in the neighborhood of Bohr's Radius, B, then we can say such balance can exist for all photons in the photon shell and that Bohr's measurements were correct for all atoms. We are treating each Unit Photon in the photon shell as a differential mass of the whole photon shell mass. By using the Unit Photon as our smallest unit of measurement, we can base all our calculations upon that sized object without needing the precise values.

We investigated force balance from two perspectives. The first is how a gravitational force may balance an electrostatic force. The second is how a magnetic force may balance an electrostatic force. Below, we present the mathematical basis for each of these possibilities.

Gravitational-Electrostatic Balance.

To begin this analysis, we re-write Eq. B3 with some variable name changes:

$$\frac{Gm_n m_t}{D^2} = \frac{Keq_n q_t}{D^2} \quad [Eq. 5]$$

Where: m_n is mass of the nucleus in kilograms (kg)
m_t is mass of one photon in kilograms (kg)
q_n is charge of the nucleus, in Coulombs (C)
q_t is charge of one photon, in Coulombs (C)
D is nucleus-to-photon shell distance, in meters (m)

At this level, the distance factor (D) once again mathematically cancels (like the range (r) did going from Eq.3 to Eq.4) and says that if we can achieve balance, that balance will occur at any nucleus-to-photon shell distance (D). The terms on the left side of the equation represent the gravitational force and those on the right side represent the electrical or electrostatic force. By inserting the values from Tables 1 and 2 into the left and right term groups, and assuming a fusion of additional proton masses into a spherical nucleus and resulting nuclear surface charges according to atomic number and atomic weight, we obtain a plot of the two forces as Figure P4-1.

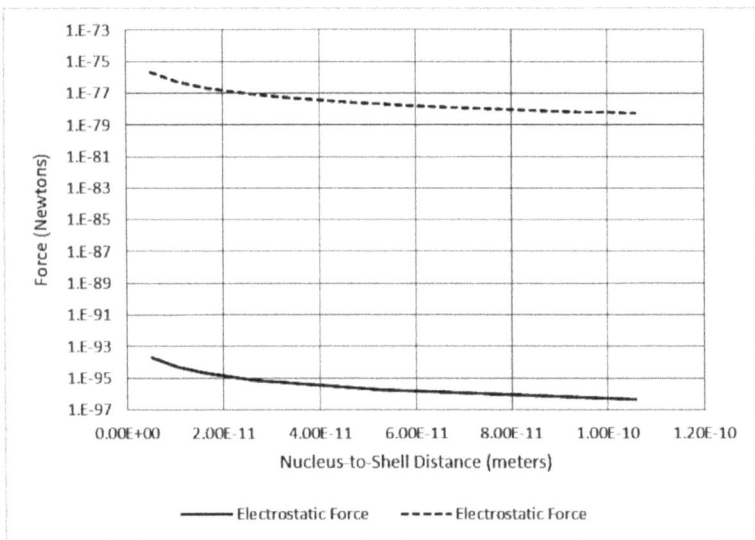

Figure P4-1. Example Force Trends for a Nominal Atomic Number = 60, Assuming Ke is Based on Permittivity of Free Space.

This shows that the electrical and gravitational forces remain proportional—never equal—at a nominal atomic number of 60. This is true for any atomic number and atomic mass, and at any nucleus-to-photon shell distance, *assuming Coulomb's constant, Ke, is true at this subatomic level.* We also can see this effect mathematically in Eq.5, since both the charge and the mass of the nucleus increase identically with added mass. (This is because in TON Particle Theory, there are no uncharged TON particle objects, equivalent to the un-charged neutrons in conventional theory).

So, if the balance is to be a function of the distance, *D*, some factor in Eq.5 has to be a function of *D*. We can use the process of elimination to isolate it. Mass or charge do not change with distance, nor does Newton's gravitational constant; these are all physical constants or non-variable particle properties. But how about Coulomb's Constant? If we look at the definition for this "constant" we see that it is actually a function of permittivity or, specifically, the permittivity of free space, as:

$$K_e = \frac{1}{4\pi\varepsilon_0} \qquad [Eq.6]$$

Where: K_e = Coulomb's electrical attraction or repulsion constant = 8.987×10^9 (N-m^2/C^2)
ε_0 = Permittivity of free space = $8.8541878176 \times 10^{-12}$ Farads per meter (F/m)

We can hypothesize that this permittivity varies as a function of the distance between particles in general, and between the nucleus and the photon shell in particular. At very close TON particle-to-TON particle distances, such as we find in the nucleus and inside the photons of the photon shell, it is the permittivity of free space (or a vacuum). At larger distances, such as the nucleus-to-photon shell distance, *D*, within an atom, we postulate that it scales inversely with distance becoming, in essence, a near perfect charge conductor (very high permittivity). To formulate and test this hypothesis, we need an expression for permittivity as a function of *D*, which we can then assess with respect to Bohr's radius. We start by isolating Coulomb's constant in a re-organized rendition of Eq.5:

$$Ke = \frac{Gm_n m_t}{q_n q_t} \qquad (Eq.5\ reorganized)$$

Next, we substitute Eq.6 into the re-organized Eq.5 above, to get an expression in terms of some intra-atomic permittivity, ε_a, instead of the permittivity of free space ε_0, to achieve force balance:

$$\frac{1}{4\pi\varepsilon_a} = \frac{Gm_n m_t}{q_n q_t} \qquad [Eq.7]$$

Re-arranging terms, we get the expression for force balance, as:

$$\varepsilon_a = \frac{q_n q_t}{4\pi G m_n m_t} \qquad [Eq.8]$$

Our next step is to formulate a relationship between the intra-atomic permittivity and the free-space permittivity that is a function, f, of the nucleus-to-photon shell distance, D:

$$\varepsilon_a = f(\varepsilon_0, D)$$

To be consistent with TON Particle Theory, the atomic permittivity must be that of free space at very close, intra-nucleus or intra-photon TON particle distances, and become extremely large at distances on the order of Bohr's radius. We believe, also, that it should vary as a function of the square of the distance. This is analogous to the effect of increasing conductivity in electrical conductors, such as copper wires, which increases with the cross-sectional area of the conductor. So, the average of the distance squared, D_{avg}, is the distance we want to use in our formulation. (This effect is consistent with Coulomb's charge theory.)

Then, the product of this average distance and some intra-atomic proportionality constant, K_a, gives us a means to achieve force balance at that distance. Hence, a possible formulation is:

$$\varepsilon_a = \varepsilon_0 + \varepsilon_0 K_a D_{avg} \qquad [Eq.9]$$

Then, since the average of some squared function, say x^2, over some range, a to b, is its integral over that range divided by the range, we have

$$x_{avg} = \frac{1}{b-a}\int_a^b x^2\, dx = \frac{x^3}{3(b-a)}$$

In this case, we are evaluating this integral to get the average value of D^2 over the range, 0 to D, so our case becomes:

$$D_{avg} = \frac{D^3}{3(D-0)} = \frac{D^2}{3}$$

And our atomic permittivity equation for force balance (Eq.9) becomes:

$$\varepsilon_a = \varepsilon_0 + \varepsilon_0 K_a \frac{D^2}{3} \qquad [Eq.\,10]$$

Now we can substitute the value of atomic permittivity from Eq.10 into Eq.8, to obtain an expression for the atomic permittivity as a function of the nucleus-to-photon shell distance, D, and that proportionality constant, K_a, as:

$$\varepsilon_0 + \varepsilon_0 K_a \frac{D^2}{3} = \frac{q_n q_t}{4\pi G m_n m_t}$$

Finally, if we use Bohr's distance, B, for D we can calculate the proportionality constant, K_a, required to achieve force balance at Bohr's distance:

$$K_a = \frac{3}{B^2}\left[\frac{q_n q_t}{\varepsilon_0 4\pi G m_n m_t} - 1\right] \qquad [Eq.\,11]$$

Now, we can experiment with our proportionality constant and see what kind of behavior we get, compared to the unbalanced force situation described by Eq.5 of this proof and associated discussions. When we calculate K_a using Eq. 11 and use it to obtain a modified permittivity (ε_a) from Eq.9, and then use ε_a instead of ε_0 to calculate K_e in Eq.6, we get a force balance situation at Bohr's distance, as expected. Using this value for K_e in Eq.2, and a range of distances (r) in Eq.1 and Eq.2, we obtain curves for electrical and gravitational force that intersect at Bohr's radius (Figure P4-2) for an atomic number of 60. The same intersection, or balance point, occurs at Bohr's distance for any other atomic number, with different curve values for the two forces.

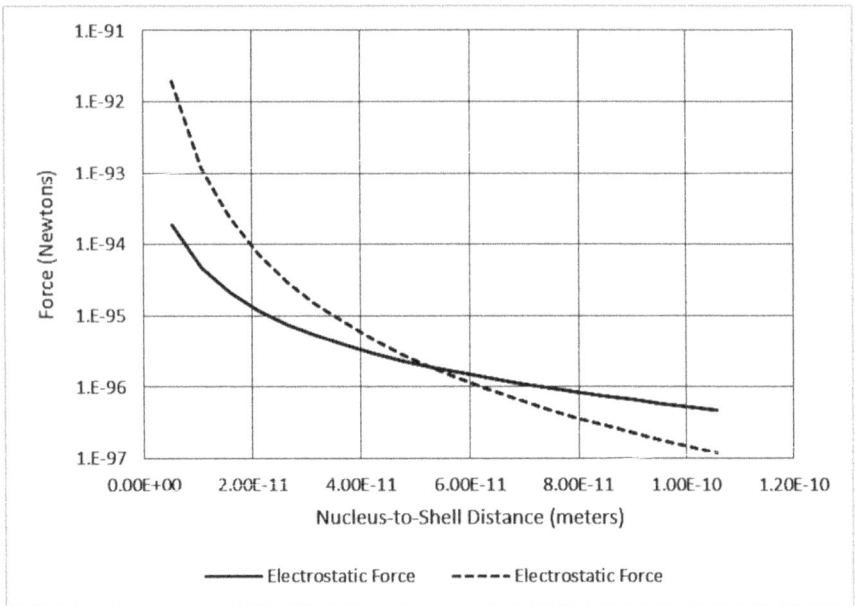

Figure P4-2. Force Balance with Variable Permittivity.

It is important to realize three fundamental assumptions in play here:

1. The intra-atomic medium can have a permittivity gradient. The electrostatic charge density diminishes outward from the surface of the proton in accordance with Coulomb's law at the incredibly small atomic distances. It is the gradient of charge density that changes the permittivity constant at these levels.[74]

2. Our calculations show that Bohr's radius is correct for the hydrogen atom. Further analysis shows that Bohr's distance is maintained throughout the periodic table due to the fact that the additional mass very slowly increases the atomic nucleus' radius. There is a very small proportional increase in the photon shell radius as the diameter of the nucleus increases. That additional distance is negligible

---

[74] In other words, it is acting like a subatomic plasma – not having a charge itself, but having the *subatomic* equivalent of variable permittivity. That may be a bit of a stretch as a plasma in conventional thought is based on ionization and non-linear dispersion of atomic (not subatomic) particles.

to the magnitude of Bohr's radius. (See Chapter 5 or Chapter 7.)

3. There is negligible magnetic attraction between the nucleus and the photon shell because the photon shell moves slowly in relationship to the atomic nucleus. There also would be negligible magnetic attraction if TON particles are not ferromagnetic.

Another issue is the consistency between TON Particle Theory and conventional atomic theory. These are the key points:

- If the photon shell is one-photon deep (in this case, one unit photon), it will have both substantially less mass and charge than a set of conventionally-defined orbiting electrons for a given atom.

- If we assign enough additional photons to the photon shell to achieve the conventionally-accepted charge of the electron set, the shell will lack the conventional electron set mass.

- If we assign enough additional photons to the photon shell to achieve the conventionally-accepted mass of the electron set, the shell will exceed the conventional electron set charge.

Adding mass to the shell will still achieve force balance in any of these cases. We would need to add additional mass to the one-TON mass we are using for TONs in the shell. In a one-TON deep shell, each TON is essentially a differential mass of the whole shell mass. To calculate for more depth and mass we would need to project the footprint of the shell inner-surface TON through the depth of the shell, thereby creating an apparent differential mass with appropriately greater mass than one TON.

Summarizing, we believe our theory of the one-TON-deep photon-populated shell is the correct one, although it is currently immeasurable by current instrumentation or technology. A physical proof will have to await invention of such measurement capabilities, but the mathematical basis for it as presented here is sound. So, in Ton Particle Theory, we simply will not expect the aggregate mass or charge of the shell for various elements to resemble the conventionally-accepted mass or charge of some number of electrons assigned to those elements.

Magnetic-Electrostatic Balance.

For this proof, we are discussing the relationship between any photon in a photon shell and an atomic nucleus as defined in TON Particle Theory. We will focus on the effects of the unit photon (two entangled TON particles spinning around each other) in a photon shell around any atomic nucleus.

To begin this analysis, we first postulate that TON particles are ferromagnetic and are therefore affected by magnetic fields. The rotational motion of charged TON particles creates an electromagnetic field perpendicular to the direction of motion and therefore exert a magnetic force upon ferromagnetic TON particles inside the atomic nucleus. This strength of this magnetic field varies in relation to the distance from nucleus to the two rotating TON particles.

To develop this proof, we can draw an analogy between a common solenoid and our two spinning TON particles (a photon). A solenoid of some cross-sectional area, a, and some number of turns, n, exerts a magnetic force, F_M, at a distance or range, D, along its central axis due

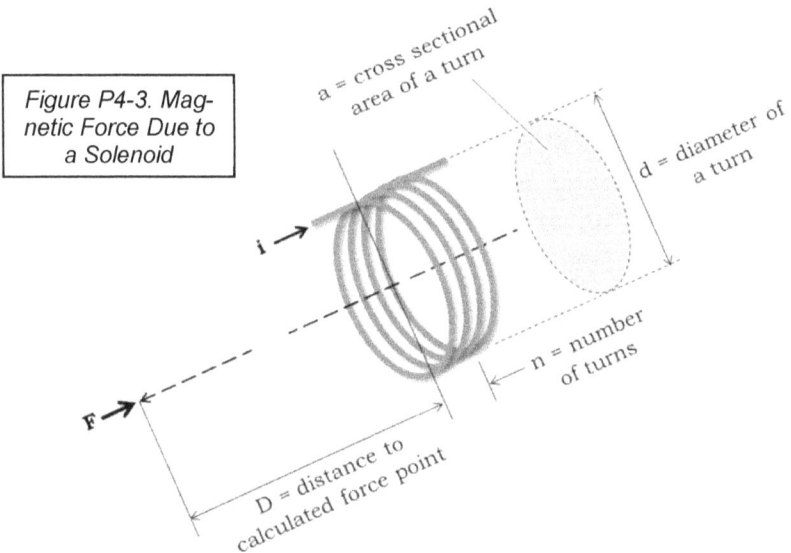

Figure P4-3. Magnetic Force Due to a Solenoid

a = cross sectional area of a turn

d = diameter of a turn

n = number of turns

D = distance to calculated force point

to the current, i, passing through it. Figure P4-3 illustrates the geometry of this situation.

The mathematical expression for the magnetic force in such a solenoid is a common working formula from electromagnetic theory, as:

$$F_M = \frac{\mu_0 (ni)^2 a}{2D^2} \qquad [Eq.\,12]$$

Where: μ_0 is the magnetic constant, or permeability of free space = 4π x 10^{-7} Henries per meter.

For our situation, we effectively have a one-turn solenoid with two charged TON particles (a unit photon) in a continuous, circular rotation of one turn. Here, the magnetic force is exerted at the distance, D, between the captured photon in the photon shell and TON particles in the atomic nucleus. So, our situation appears as depicted in Figure P4-4.

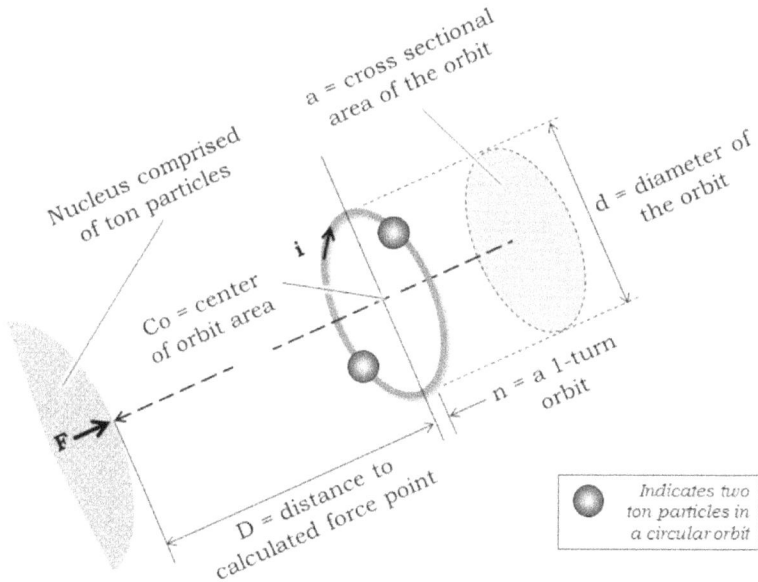

Figure P4-4. Magnetic force at the Nucleus Due to a Pair of Orbiting TON Particles (a unit photon)

According to Maxwell's Law, moving charges (TON particles) create an electrical current, or the flow of charge in a circular path, that occurs in some amount of time, t, creating the current, i, as:

$$i = \frac{q}{t} \qquad [Eq.\,13]$$

Here, the time, t, is a function of the velocity of the particles, v, and the distance they travel in their rotation, which represents the circumference of the solenoid area, a. So, if a has some diameter, d, the velocity, v, of these particles is the circular distance traveled divided by the time, t, as:

$$v = \frac{\pi d}{t} \quad or \quad t = \frac{\pi d}{v} \qquad [Eq.\,14]$$

Then, substituting Eq.14 into Eq.13 gives us an expression for current as a function of charge, velocity, and rotational diameter, as:

$$i = \frac{qv}{\pi d} \qquad [Eq.\,15]$$

Now we can substitute Eq.15 into Eq.12, and set n = 1 for our one-turn solenoid-like rotation, to obtain an expression for force as a function of charge, rotational velocity, and axial distance, as:

$$F_M = \frac{\mu(qv/\pi d)^2 a}{2D^2} \qquad [Eq.\,16]$$

But the solenoid cross-sectional area, a, can also be expressed in terms of its diameter $(a = \pi(d/2)^2 = \pi d^2/4)$ which allows a simplification of Eq.16, as:

$$F_M = \frac{\mu(qv/\pi d)^2 \pi d^2}{2D^2(4)} = \frac{\mu(qv)^2}{8\pi r^2} \qquad [Eq.\,17]$$

Next, the general expression for charge, q, may be expressed as the product of the two rotating particle charges $(q = 2q_p)$, as:

$$F_M = \frac{\mu(2q_p v)^2}{8\pi D^2} \qquad [Eq.\,18]$$

Finally, we can factor out the magnetic constant (permittivity of free space) and separate some variable products, which gives us a final expression for the force as a function of particle charge, orbital velocity, and axial distance, as:

$$F_M = \frac{(4\pi)10^{-7}(2q_p v)^2}{8\pi D^2} = \frac{2q_p^2 v^2 \; 10^{-7}}{D^2} \qquad [Eq.\,19]$$

Now, if we hypothesize a situation where the magnetic force as defined above is in a direction to oppose and cancel out the electrostatic force due to Coulomb's law on a differential area basis, we have a balance of force, as:

$$F_M = 2F_E \qquad [Eq.\,20]$$

Here, we have multiplied the electrical force by two because the magnetic force is due to *two* orbiting TON particles (a unit photon), so it must be balanced, at a differential area basis, by forces between those two TON particles.

Then, substituting Eq.19 and Proof1, Eq.2 into Eq.20, the balance equation becomes:

$$\frac{2q_p^2 v^2 \; 10^{-7}}{D^2} = \frac{2Keq_p^2}{D^2} \qquad [Eq.\,21]$$

But Coulomb's constant, *Ke*, at the permittivity of free space, can be stated as a function of the speed of light, *c*, as:

$$Ke = c^2 \; 10^{-7}$$

So, substituting this expression for *Ke* into Eq. 21, our final force balance equation becomes:

$$\frac{2q_p^2 v^2 \; 10^{-7}}{D^2} = \frac{2q_p^2 c^2 \; 10^{-7}}{D^2} \qquad [Eq.\,22]$$

And we see that all factors except the particle velocity and speed of light cancel out. Significantly, this says that balance is *independent of distance*, as it was for the gravitational-electrostatic balance discussed earlier in this proof. So, the ratio of electrical force to magnetic force is:

$$\frac{2F_E}{F_M} = \frac{c^2}{v^2} \qquad [Eq.\,23]$$

Now, applying the specifics of TON Particle Theory, this equation establishes some specific principles, on a differential area basis, where the differential areas are those (a) occupied by two TON particles on the surface of the nucleus and (b) the two rotating TON particles, or unit photon, on the surrounding photon shell. We describe those specific principles below.

First, we might speculate that the captured photons in the photon shell are affected by more than just the TON particles on the atomic nucleus' surface for both the electrical and magnetic force calculations. This does not affect the balance defined in Eq.23. As Figure P4-5 illustrates, we could define a spherical conical volume, Vc, defined by the intersection of vectors from the shell orbit extremities to the nucleus center (conical section A'-Cn-B' in the figure), or a full or half spherical volume, Vs, for the whole nucleus (sphere or hemisphere A-Cn-B in the figure), or any other volume of the nucleus we like (Va, for "anything").

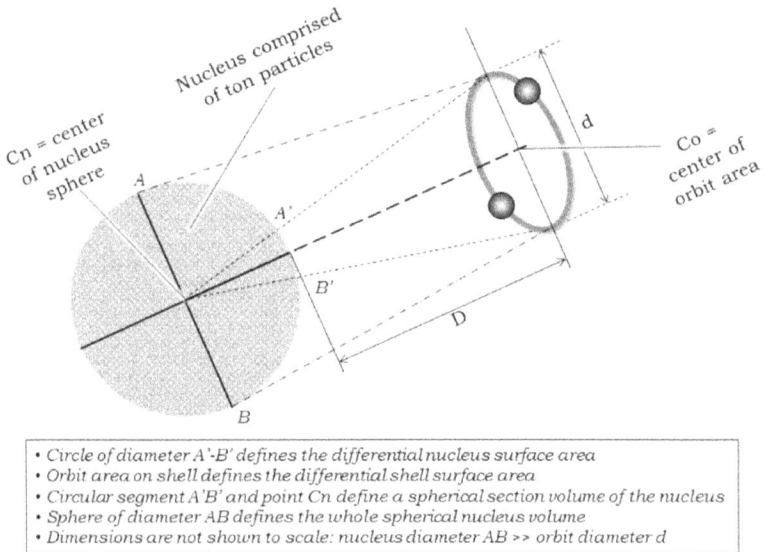

• Circle of diameter A'-B' defines the differential nucleus surface area
• Orbit area on shell defines the differential shell surface area
• Circular segment A'B' and point Cn define a spherical section volume of the nucleus
• Sphere of diameter AB defines the whole spherical nucleus volume
• Dimensions are not shown to scale: nucleus diameter AB >> orbit diameter d

Figure P4-5. Volumetric Variations for the Force Balance Equation.

Here, the forces would be the volume integral (3-dimensional triple-integral) of the differential forces over the defined volume), so for such cases, the force relationship of Eq.23 becomes

$$\frac{\int_{Vc} 2F_E}{\int_{Vc} 2F_M} = \frac{\int_{Vs} 2F_E}{\int_{Vs} 2F_M} = \frac{\int_{Va} 2F_E}{\int_{Va} 2F_M} = \frac{2F_E}{F_M}\frac{c^2}{v^2} \qquad [Eq.24]$$

Basically, at the differential area level as discussed above, we are saying that every pair of rotating TON particles (unit photon) in the photon shell is balanced against every two TON particles on the nucleus' surface; and this effect can be projected by mathematical integration to relate rotating TON particle pairs in the photon shell to TON particles throughout the volume of the atomic nucleus. Therefore, the force balance concept still holds true. Given these force balance relationships, we can turn to the issue of photon shell diameter.

First, we can calculate the diameter of the nucleus, d_n, from a spherical volume that contains the number of TON particles that need to accumulate in order achieve the mass equivalent of a conventional nucleus, as explained in Proof 1. Next, we calculate the surface area of the nucleus, a_n, using conventional geometry as:

$$a_n = \pi d_n^2 \qquad [Eq.\,24]$$

Then we determine the number of TON particles on the nucleus surface, N, as a function of the nucleus surface area, the cross-sectional area of a TON particle, d_t, and our area packing factor (discussed in Proof 1), F_a, as:

$$N = \frac{F_a a_n}{\pi d_t^2} \qquad [Eq.\,25]$$

Due to the force balance relationships discussed earlier, we know that the number of TON particle objects (two per photon) in the photon shell needed to balance the TON particles in the nucleus surface is that same number, N. Hence, with two TON particles per rotation, there must be $N/2$ rotating pairs (photons) or circular areas.

Now, consider a pair of spinning TON particles (a photon) in the shell. As Figure P4-6 shows, the diameter of rotation area, a_o, must be slightly larger than the diameter of the two spinning TON particles, d_t to avoid either of them from colliding with the other.

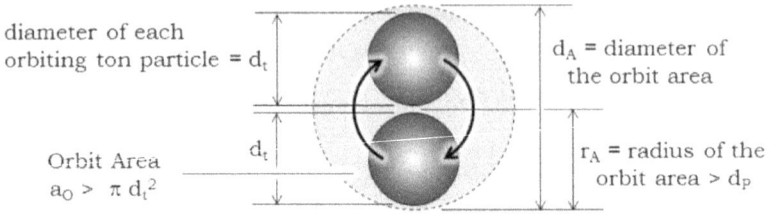

Figure P4-6. Area and Radius Relationships in the Orbiting TON Particle Pair of a Unit Photon

So, the radius of the orbit area, r_o, is always slightly larger than the TON particle diameter, and so the orbit area, a_o, is always slightly greater than an area calculated using TON diameter as the area radius, as:

$$a_o = \pi r_o^2 = \pi d_t^2 \qquad [Eq.\,26]$$

Next, since the surface area of the photon shell must be large enough to encompass $N/2$ orbit areas, a_o, the surface area of the shell, a_s, is a function of these quantities and the photon shell packing factor, as:

$$a_s = (N/2)a_o/F_a \qquad [Eq.\,27]$$

And we can calculate the *minimum possible diameter* of the shell, d_s, from the summation of the photon shell surface as:

$$d_s = \sqrt{\left(\frac{a_s}{\pi}\right)} \qquad [Eq.\,28]$$

Due to how the packing factors affect the numbers of TON particles in the nucleus and in the photon shell, the minimum diameter of the photon shell is slightly larger than the diameter of the nucleus, since the nucleus' surface TON particles pack more tightly than the captured photons in the photon shell, which must include enough space to avoid collisions as Figure P4-7 shows.

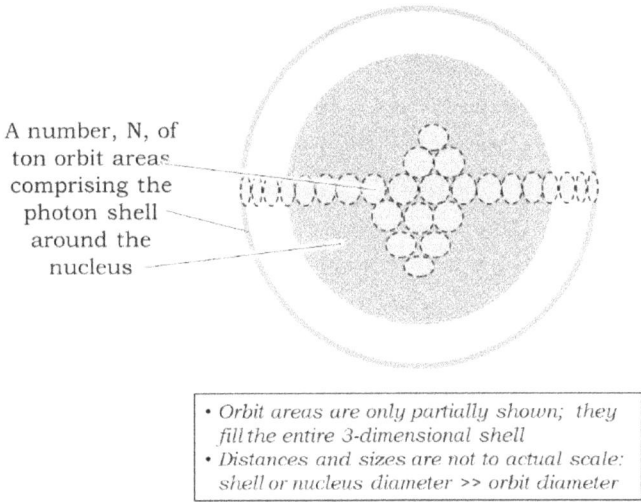

A number, N, of ton orbit areas comprising the photon shell around the nucleus

- *Orbit areas are only partially shown; they fill the entire 3-dimensional shell*
- *Distances and sizes are not to actual scale: shell or nucleus diameter >> orbit diameter*

Figure P4-7. Graphical Rendition of the Shell Composition of Photons

Figure P4-8 (below) shows how the photon shell radius and nucleus radius mathematically relate over the range of elemental atomic numbers. The minimum possible shell radius is larger than the proton diameter by the square root of two since each photon's rotational area is at least twice the TON particle cross-sectional area.

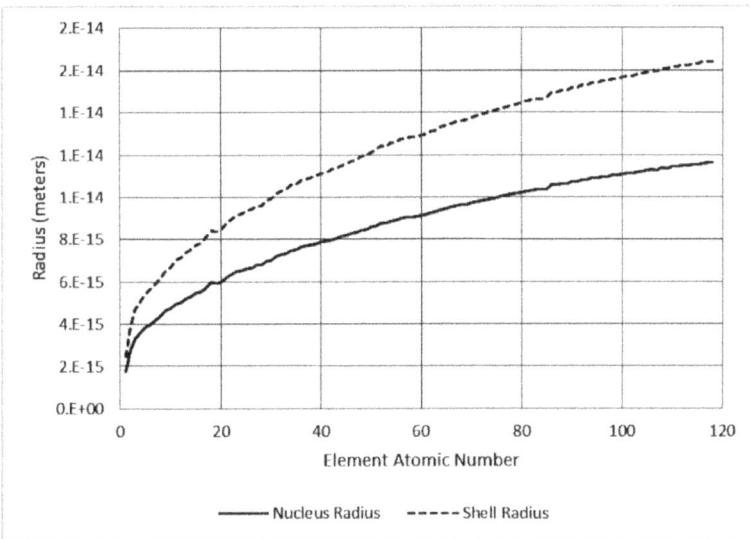

Figure P4-8. Comparison of the Minimum Possible Shell and Nucleus Radii at Various Atomic Numbers

As discussed above, these photon shell dimensions represent the minimum possible photon shell radius, or distance between the nucleus and the photon shell. Since the electromagnetic-to electrostatic balance is independent of distance, as the earlier mathematics showed, that distance, or photon shell radius, may be any reasonable value. For example, if we assume a photon shell diameter 21843 times larger than the minimum TON particle rotation diameter, we get the photon shell radii of Figure P4-9, which range from Bohr's distance for hydrogen to about six times Bohr's distance at the highest atomic number. So, Bohr's theory is completely plausible within this force balance concept.

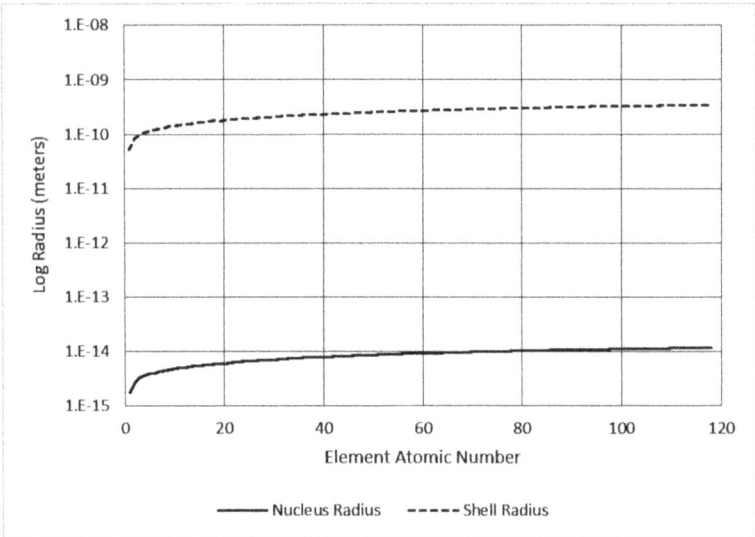

Figure P4-9. Creation of Shell Radii Near Bohr's Distance at Various Atomic Numbers

As a corollary to this proof, by substituting the rotational area expression of Eq.26 into Eq.27, we have:

$$\pi d_n^2 = N\pi d_t^2/F_a \ or \ d_n^2 = N d_t^2/F_a \qquad [Eq.\,29$$

And we can re-arrange Eq.29 to calculate N, the number of rotational areas, or pairs of spinning TON particles in the photon shell, as:

$$N = F_a d_n^2 / d_t^2 \qquad [Eq.\,30]$$

Since N is the number of spinning TON particle pairs, the number of TON particles in the photon shell is *2N*, which can be multiplied by the charge of a TON particle or the mass of a TON particle to determine the mass and charge of the entire photon shell. This is mathematically the same as the number of TON particles in a one-TON particle-deep surface area of the nucleus, due to our earlier assumption of TON particle-to-TON particle equivalence in a differential area.

Calculating the photon shell mass and charge using the equations just described produces charge and mass quantities that are significantly smaller than those of a conventionally-defined "electron," by factors of tens of orders of magnitude. As we said in the introduction to these proofs, TON Particle Theory's photon shell is the accumulation of photons surrounding the nucleus, and bears no resemblance to a conventionally-defined "electron particle."

Summarizing, this proof established the feasibility of an electrostatic-to-electromagnetic force balance that could exist at Bohr's distance, or other distances for that matter, for a photon shell that is the fundamental part of TON Particle Theory atomic model. We have established that you can have a photon shell around an atomic nucleus, and that photon shell will retain its distance to the nucleus regardless of the atomic nucleus. This means the TON Particle Theory atomic model encompasses all known atomic elements.

Proof 5: Atomic Stability in TON Particle Theory

Earlier in this book (Chapter 7), we introduced the idea that an irregularly formed (non-spherical) atomic nucleus would cause its photon shell to be unstable. We hypothesized that there was a certain number of equivalent individual protons in an atomic nucleus that would then be "spherical enough" to support a stable photon shell.

To look at this issue, we created an instability factor variable that we could use to assess the relative stability of an accumulation of un-fused, individual proton-sized masses in an atomic nucleus. These unfused mass objects create geometrical nuclei that would require more volume than a single spherical object with the equivalent mass. As we increase the number of objects, the overall geometry converges on a more sphere-like object, but will still require significantly more volume than a single sphere of equivalent mass. Hence, we wanted our instability factor to be equal to 1.0 for a single spherical proton or for a mathematically hypothetical infinite number of proton-sized objects, which would make the resulting mass almost perfectly spherical simply due to its huge volume.

In Proof 4, we demonstrated that a photon shell is mathematically supported by single spherical nuclei. In this proof, we will look at the effect of an irregularly shaped nuclei based on proton-sized objects to verify that the resulting geometry causes an instability in that photon shell due to minute imbalances of the forces involved.

To illustrate this instability effect, we look at the case where we establish a sphere composed of just a few spherical objects, compared to a sphere of more spherical objects (Figure P5-1).

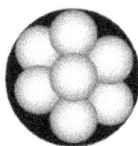

Figure P5-1. Cluster of Photon Spheres in a Nucleus

Conjoined mass of a few proton spheres

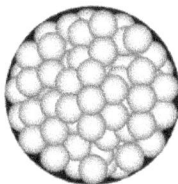

Conjoined mass of many proton spheres

Looking at the sphere on the left, you have a much more irregular surface area than in the sphere on the right. This is apparent from the black area around the surface, that the resulting electrostatic force vectors would vary greatly across the surface of the sphere. These variations will establish areas of force imbalance along the under surface of the surrounding photon shell. As we established in Proof 4, these changes in permeability have a definable effect on each captured photon. Looking at the figure on the right, we see significantly less black area along the surface, which indicates that there would be a more consistent level of electrostatic force around the sphere, thus creating a more consistent level of permeability at the photon shell. Therefore, we can see it is this outer layer of protons that has the most effect on the photon shell, so that the more proton equivalents in the outer nucleus surface determines the overall photon shell stability.

We approached this instability concept mathematically, using a heuristic methodology we describe below. We can use spherical volume in the ensuing calculations since density multiplied by volume equals mass. We use a one proton-sized spherical object as our "unit volume" in these calculations, which makes Na = atomic mass number. For example, a nucleus with 17 protons has an accumulated proton volume of Na = 17.

Gauss and others estimated and mathematically verified that the maximum density packing factor, Fp, for a spherical cluster of same-size spheres is calculated as Fp = $\pi/(3(20.5)) = 0.74048$. This occurs, as a minimum situation, when a sphere is surrounded by 12 identical spheres. So, the packed volume, Vp, of Na protons is the number of spheres in the cluster divided by the packing factor, as:

$$Vp = Na/(Fp) = Na/0.74048$$

This packed volume represents the total volume of the spheres and the spaces in-between them, as defined by the packing factor. The packed volume, although not a perfect sphere, is sphere-like. For now, we'll ignore any instability-wrought surface irregularity issues and treat this packed volume as a sphere. The volume of a sphere, Vs, is related to its diameter, Ds, as:

$$Vs = \pi(Ds^3)/6$$

Or by re-arrangement of terms as:

$$Ds = (6Vs/\pi)^{1/3}$$

So, the diameter of our packed spherical volume, Dp, is calculated as:

$$Dp = (6Vp/\pi)^{1/3}$$

As mentioned above, when the proton spheres are touching (not fused), this spherical mass (with volume = Vp and diameter = Dp) is not a true, smooth sphere, since it is an aggregation of many smaller spheres. It's a "bumpy" sphere cluster. For calculation and analysis purposes, we divide the spherical mass into two regions, a central or "core" volume and a three-dimensional surface layer volume whose radial depth is nominally about one proton deep, in any radial direction (from the surface to the center of the packed volume). Our aim was to interpret these core and surface layer volumes to develop a heuristic atomic instability factor. Figure P5-2 illustrates the concept of core and surface volumes.

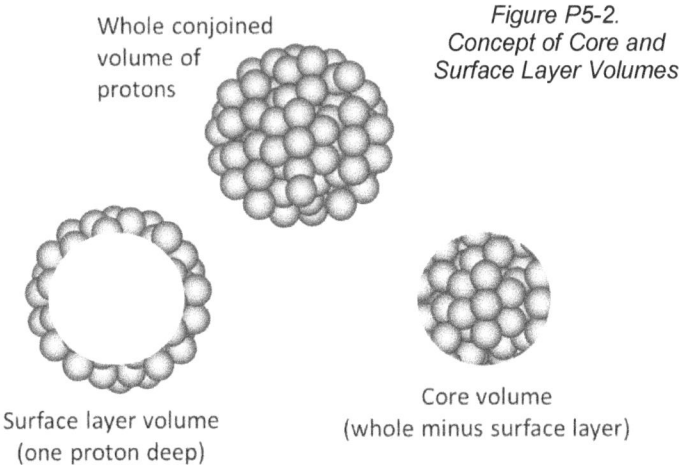

Whole conjoined volume of protons

Figure P5-2.
Concept of Core and
Surface Layer Volumes

Surface layer volume
(one proton deep)

Core volume
(whole minus surface layer)

Since we have defined the surface layer as being about one proton in depth, the core sphere diameter of the packed cluster, *Dc*, must be about two proton sphere diameters less than *Dp*, which is the packed cluster diameter. We'll return to this diameter value and quantify it more explicitly later. But for now, we calculate the diameter of our unit spherical volume, Du, with a volume of 1.0 (proton volume) as:

$$Du = (6/\pi)^{1/3} = 1.2407 \text{ (linear measure units)}$$

And we calculate the diameter, Dc, of the core volume within the packed volume, Vp, as:

$$Dc = Dp\text{-}2(1.2407) \text{ (linear measure units)}$$

Now we can calculate the volume of the packed core region, Vcp, as:

$$Vcp = \pi(Dc^3)/6$$

We assume that the core has a tightly packed arrangement according to Gauss' packing factor, Fp.[75] We can calculate the accumulated core volume, Vca, simply by multiplying the packed core volume, Vcp by Gauss' packing factor, Fp, as:

$$Vca = Vcp(Fp)$$

Next, we want to calculate the number of proton spheres in the core volume. Since we are dealing with touching and not fused proton masses or spheres, we will not allow a fraction of a sphere in our core volume sphere count, which we calculate as:

$$Nc = \lfloor Vca \rfloor$$

Where $\lfloor x \rfloor$ indicates a floor function; truncates the fractional part of x.

At this point we calculate the estimated number of proton spheres in the surface layer, Ns, as the number in the core, Nc, subtracted from the number in the accumulated mass, Na, as:

$$Ns = Na - Nc$$

Our hypothesis was that we could interpret this number of spheres in the surface layer to derive an instability factor for the nucleus. If we simply divide the number of spheres in the surface layer

---

[75] However, looking forward a bit, we are going to use the core volume and surface layer volumes to derive an instability factor. So, we want our core volume to represent *accumulated* unit spheres, and not *packed* unit spheres.

volume by the number of spheres in the whole aggregated cluster, we get a fraction representing the surface layer sphere count relative to the whole cluster sphere count, which is a measure of the surface layer irregularity graphically depicted earlier in Figure P5-1. However, we will by definition always have at least one proton sphere in the core as long as the total ≥ 12, so we diminish the number of surface layer spheres by one and calculate an intermediate relative, normalized stability (not instability) factor, Sr, as:

$$Sr = 1 - (Ns-1)/Na$$

As more and more spheres accumulate, the whole volume sphere count, Na, increases at a higher rate than the surface layer sphere count, Ns. Then, by dividing Sr into unity (1.0), we get a normalized relative instability factor, Ir, as

$$Ir = 1/Sr$$

As we discussed earlier, there are two defined values for such an instability factor, at Na = 1 (where Ns must be 0) and the limit of Ir as Na approaches infinity (∞). Where Na = 1, there is only one sphere, and the aggregation is ideally stable; so we expect, and see, that Ir = 1. This means that the instability is one times the instability of a solid, unit-volume sphere, or ideally stable. Where Na approaches infinity, the (Ns-1)/Na term approaches zero, since the surface volume becomes infinitely smaller than the whole aggregation. Here, since Ir = 1 - (Ns-1)/∞, it follows that Ir = 1, and once again, we have an ideally stable cluster or aggregation of spheres. Figure P5-3 plots this instability factor as a function of the number of spheres or proton volumes accumulated over a reasonable range of atomic mass numbers, and Figure P5-4 plots the same data in logarithmic values of proton volumes extended to 1,000,000, showing the convergence to 1.0 at the endpoints.

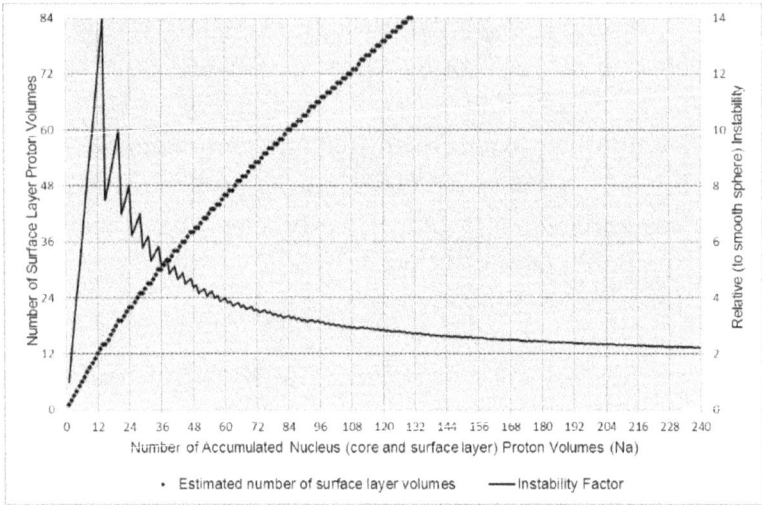

Figure P5-3. Instability Factor (Ir) as a Function of Accumulated Volumes

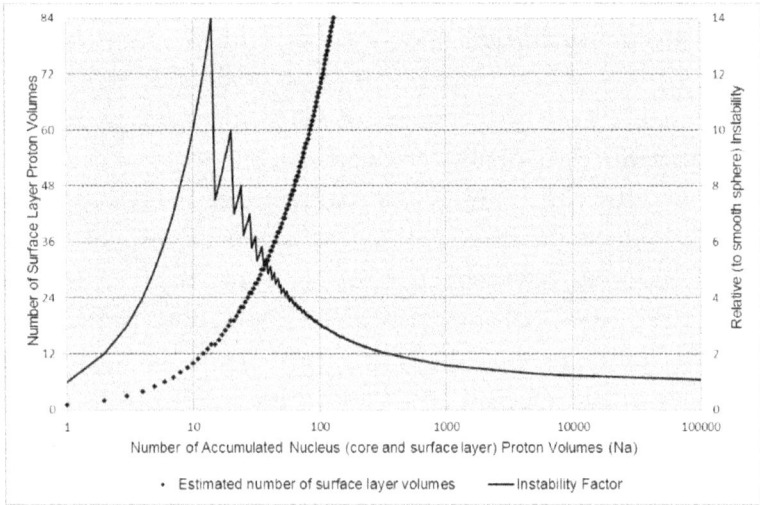

Figure P5-4. Instability Factor (Ir) as a Function of Accumulated Volumes (log scale)

At this point, let us return to the estimation of the surface layer depth dimension, which we estimated earlier as one proton volume diameter = 1.2407 linear measure units. This depth estimation was probably too high – a maximum value, since the proton spheres in the surface layer would be fitting into the irregular surface of the core volume at the surface-core volume boundary. We do not have a quantified way to estimate this value under our rather heuristic approach. However, if we assume a reduction in that maximum value to a value of 0.9, we get an interesting effect. Here, as Figure P5-5 shows, the mathematics calculates a maximum instability factor of 12 when the accumulated mass consists of 12 proton volumes. Adding one more proton mass to the accumulation reduces the instability factor significantly, when the proton volume count is 13 and this (13) is the minimum number of like-sized spheres, which exactly satisfies Gauss' packing factor.

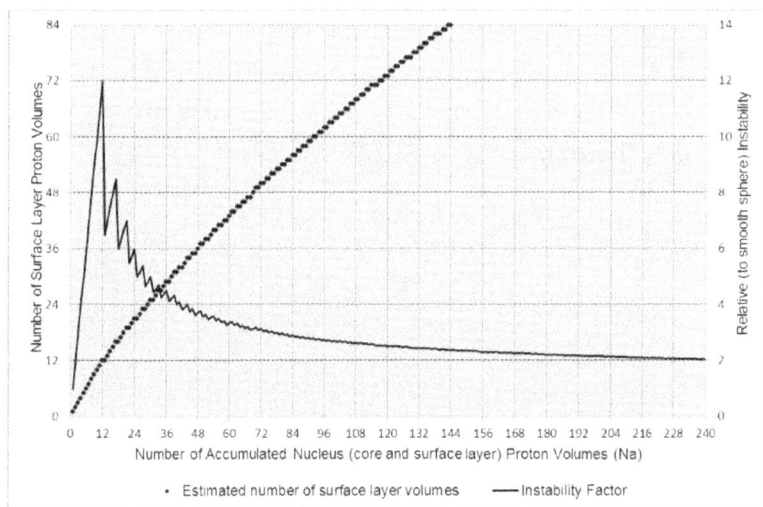

Figure P5-5. Instability Factor (Ir) as a Function of Accumulated Volumes with More Appropriate Surface Layer Depth

There are a couple of ways to rationalize the 0.9 value. One is that multiplying the calculated proton diameter of 1.2407 by the Gaussian packing factor (0.74) gives a value approximately equal to 0.9. Another is that 0.9 approximates the cubed root of the packing factor (Fp1/3 = ~0.9), suggesting a radial component of the packing factor. It would be interesting to test our heuristic concepts using some actual 3-dimentional volume packing modeling, but we know the maximum value for this depth is one proton diameter, and the actual depth is a function of how efficiently all these proton spheres packed together.

The fact that using a surface layer depth = 0.9 produces (a) conver-gence to 1.0 at the endpoints and (b) calculates a core count of 1.0 when the proton volume count reaches 13, are strong indicators that this is an accurate measure, in the context of our applied heuristics.

Summarizing, the appropriate interpretation of the instability fac-tor, Ir, is as a multiplier relative to the stability of a single sphere or a fused mass of spheres. So, for example, we see in Figure P5-5 that Ir = 4 when Na = 42, which says that, when forty-two proton masses are touching, this assembly is four times more unstable than when forty-two proton masses are fused into a single sphere.

In any atomic nucleus of a mass less than thirteen proton equiv-alents, the atom could not form a stable photon shell unless that nu-cleus were a single, smooth, fused sphere. Since we know that such atoms exist, we find this evidence greatly supports our atomic model over the classic model.

About the Authors

Donald A. Bertke began his career as an electronics technician. He earned a Bachelor of Science in computer engineering while working full-time and later earned a Master of Science in computer science. His work experience included lasers, spectroscopy, programming the first microprocessor-controlled laser-guided bomb (LGB), search and rescue systems, integrated avionics, weapons of mass destruction (WMD) analysis, information systems, and created a full Capability Maturity Model Integrated (CMMI) compliant training program for systems and software engineering.

His work has covered nearly all science domains at very high levels of detail. All these experiences led to creating the solution presented in this book.

Past papers include topics on system and network timing analysis and a patent application on an adaptive algorithm. He is also co-author on the *World War II Sea War* series that documents detailed naval activities for most ships during that war. This book expands on his first technical effort since being disabled in 2001.

Herbert L. Hirsch is a senior consulting engineer with a B.S. in electrical engineering from the University of Cincinnati. He has forty years of experience designing electronic systems, system and phenomenology simulations, and signal/image processing algorithms primarily for military sensor systems. Mr. Hirsch began his career at Systems Research Laboratories in Dayton, Ohio, where he held engineer and senior engineer positions. He was an engineering group leader for Quest Research Corporation, the director of systems engineering for Applications Research Corporation, the director of engineering for Defense Technology, Inc., and chief engineer for MTL Systems, Inc.—all in Dayton, Ohio. He currently operates his consulting business, Hirsch Engineering and Communications, Inc., from his home in Vandalia, Ohio.

Mr. Hirsch has published or co-published numerous books on

communication, radar antenna simulation, statistical signal processing, hardware description languages, and electronic counter-
measures, including The Essence of Technical Communication for Engineers (IEEE Press, 2000), Effective Design and Testing with WAVES
and VHDL (Kluwer Academic Publishers, 1996), Chapter 11 in The
CM Handbook for Aircraft Survivability (USAF, 1994), Statistical Signal Characterization (Artech House, 1991), and Practical Simulation
of Radar Antennas and Radomes (Artech House, 1988).